Supporting Expeditionary Aerospace Forces

Expanded Analysis of LANTIRN Options

Amatzia Feinberg

Hyman L. Shulman

Louis W. Miller

Robert S. Tripp

Prepared for the UNITED STATES AIR FORCE

Approved for public release; distribution unlimited

Project AIR FORCE RAND

The research reported here was sponsored by the United States Air Force under Contract F49642-96-C-0001. Further information may be obtained from the Strategic Planning Division, Directorate of Plans, Hq USAF.

Library of Congress Cataloging-in-Publication Data

Supporting expeditionary aerospace forces : expanded analysis of LANTIRN options / Amatzia Feinberg ... et at.
p. cm.
"MR-1225-AF."
Includes bibliographical references.
ISBN 0-8330-2903-7
1. LANTIRN (Military aeronautics)—Maintenance and repair. 2. United States.
Air Force—Equipment—Maintenance and repair. I. Feinberg, Amatzia, 1966– II. Rand Corporation.

UG1103 .S88 2000
358.4'1422—dc21

00-064026

Published 2001 by RAND
1700 Main Street, P.O. Box 2138, Santa Monica, CA 90407-2138
201 North Craig Street, Suite 102, Pittsburgh, PA 15213-1516
1200 South Hayes Street, Arlington, VA 22202-5050
RAND URL: http://www.rand.org/
To order RAND documents or to obtain additional information, contact Distribution Services: Telephone: (310) 451-7002;
Fax: (310) 451-6915; Internet: order@rand.org

PREFACE

Although much work remains to define and prepare Air Force units for Expeditionary Aerospace Force (EAF) responsibilities, it is clear that EAF concepts will play a central role in the future Air Force. The EAF relies on rapidly deployable, immediately employable, and highly flexible forces to serve a strategic role as an alternative to a large permanent forward presence in deterring and responding to aggressive acts. EAF success will, to a great extent, depend on the effectiveness and efficiency of the system supporting flying operations. The Air Force has named such a support system one of its six necessary core competencies and labeled it the Agile Combat Support (ACS) system.

ACS efficiency and effectiveness are affected by decisions made across planning, programming, and budgeting system timelines. Far-term ACS decisions affect support structures required to meet future operational requirements. Mid-term ACS decisions affect the design, development, and evolution of the support infrastructure for meeting operational requirements within the programming and budgeting time horizons. Near-term decisions affect where, when, and how existing resources are employed. Across this time spectrum, logistics requirements can be satisfied in a variety of ways, each with different costs, flexibility, response times, and risks. This study addresses logistics structure alternatives for meeting demands for Low Altitude Navigation and Targeting Infrared for Night (LANTIRN) across the spectrum of EAF operational requirements from major theater wars to peacetime operations and amends earlier RAND research with new data collected during the Air War Over Serbia (AWOS).

In this study we compare the current decentralized policy, in which intermediate maintenance capabilities are deployed with flying units, with consolidated options in which maintenance capabilities do not deploy. This dipole decision space offers many opportunities while introducing multiple risks—all of which the Air Force must consider.

This research, sponsored by the Air Force Deputy Chief of Staff for Installations and Logistics (AF/IL), was conducted in the Resource Management Program of Project AIR FORCE. This and related projects seek to identify ways of enhancing the effectiveness of the Air Expeditionary Force.

This report should be of interest to logisticians, operators, and mobility planners throughout the Air Force. It is one of a series addressing ACS options to enhance the effectiveness of EAF operations. Other titles in this series include *An Integrated Strategic Agile Combat Support Planning Framework* (MR-1056-AF, 1999), *New Agile Combat Support Postures* (MR-1075-AF, 2000), *Flexbasing: Achieving Global Presence for Expeditionary Aerospace Forces* (MR-1113-AF, 2000), *An Analysis of F-15 Avionics Options* (MR-1174-AF, 2000), and *A Concept for Evolving the Agile Combat Support/Mobility System of the Future* (MR-1179-AF, 2000).

PROJECT AIR FORCE

Project AIR FORCE, a division of RAND, is the Air Force Federally Funded Research and Development Center (FFRDC) for studies and analyses. It provides the Air Force with independent analyses of policy alternatives affecting the development, employment, combat readiness, and support of current and future aerospace forces. Research is performed in four programs: Aerospace Force Development; Manpower, Personnel, and Training; Resource Management; and Strategy and Doctrine.

CONTENTS

FIGURES

SUMMARY

SUPPORT OPTIONS FOR A NEW PARADIGM

> *We have moved away from a containment strategy to one of global engagement with shaping and responding as the key words for the United States Air Force.*[1]

> *The increasing number of deployments launched on short notice to unpredictable locations presents new challenges to Air Force personnel and capabilities.*[2]

This paradigm shift presents new challenges to legacy support structures and the evolving Agile Combat Support (ACS) system. Support must spin up to sustain operations almost immediately, minimize airlift demands to increase deployment speed, and have the flexibility to respond to uncertain locations and mission requirements. Concurrently, cost pressures and the personnel considerations of an expeditionary force have led the Air Force to reexamine the complete ACS system to understand how alternative structures, technologies, and methods affect capabilities.

---

[1]General Michael E. Ryan, USAF, "Air Expeditionary Forces," Department of Defense press briefing, Washington, D.C., August 4, 1998.

[2]General Michael E. Ryan, USAF, "Aerospace Expeditionary Force: Better Use of Aerospace Power for the 21st Century," briefing to USAF (HQ), Washington, D.C., 1998.

This report examines alternative Low Altitude Navigation Targeting Infrared for Night (LANTIRN) intermediate maintenance operations and explores the implications of support equipment investments in conjunction with various logistics concepts. This study builds on a series of RAND research projects on the evolving ACS system and considers the implications of decisions based on capabilities rather than costs. We address recurring labor and transportation costs; investment costs associated with the options assessed were not available at the time of this study.

The LANTIRN system consists of two pods (navigation and targeting) employed by F-16s and F-15Es. The alternative support structure options range from the current decentralized practice of deploying intermediate maintenance with the fighting units to a network of consolidated support locations (or even a single location). Support equipment upgrades, policies, and capabilities combine with these structure options to form a rich array of possibilities from which the Air Force can choose the best ACS system to meet uncertain scenarios. Our goal is to highlight the key issues affecting these alternatives and to illustrate some of the tradeoffs the Air Force faces in making these decisions.

We consider various investment options and probe assumption sensitivities across several operational scenarios. Scenarios include an extended illustrative Time Phased Force and Deployment Data (TPFDD) deployment program and a halt phase employment scenario. We investigate Line Replaceable Unit (LRU) repair system requirements as well as other systems potentially used for precision attack (LITENING II[3]). Among the assumption sensitivities we probe are those on personnel, productivity, transportation time, and equipment availability. We also examine the finances of each option.

Our research shows that consolidating the LANTIRN intermediate maintenance support system may offer advantages in enhancing operational flexibility, improving support responsiveness, and decreas-

---

[3]LITENING II is a targeting pod with capabilities similar to the LANTIRN targeting pod, but it relies on a two-level maintenance system whereas LANTIRN's is a three-level. Existing LANTIRN-capable aircraft would require modification to use LITENING II. The Air National Guard is planning to purchase LITENING II pods to augment its precision attack capabilities.

ing the requirements for highly skilled personnel. A regional support structure, however, would be more sensitive to transportation delays and require greater cross-organizational communication. More important, data collected in the Air War Over Serbia (AWOS) revealed that the USAF may fall short on support equipment by at least 50 percent if it is to meet the requirements of two coincident major theater wars (MTWs).

SCENARIOS, SUPPORT STRUCTURES, AND EQUIPMENT UPGRADES CREATE THE TRADE SPACE

The Air Force currently maintains LANTIRN pods using a decentralized logistics structure, deploying full sets of testers from home operating bases to forward operating locations (FOLs) with the aircraft. Other options rely on varying levels of consolidation, ranging from using a single CONUS support location (CSL) to using a CSL in network with two to four forward support locations (FSLs). This analysis centers on the implications of various levels of consolidation chosen for the LANTIRN intermediate-level support operations relative to operational scenarios ranging from peacetime to two coincident MTWs. We focus on consolidated versus decentralized support and highlight our findings based on data collected in the AWOS. Specifically, we were able to assess illustrative wartime removal rates and transportation times more accurately with the new data.

Structure decisions may concentrate on support locations, but they should not do so exclusively. Adopting new procedures or technologies can affect how various support structures compare with each other in terms of capabilities and costs. Although the Air Force does not plan on upgrading pod performance or purchasing additional LANTIRN pods, we evaluated three investment options to upgrade the current support equipment (LANTIRN Mobility Shelter Set—LMSS) used to repair these pods: "zero" investment, Advanced Deployment Kit (ADK) and Mid-Life Upgrade (MLU). The upgrades offer a reduced footprint (the amount of initial airlift space needed to transport operating materiel and combat equipment) and potentially enhanced support equipment performance and reliability.

Combining scenarios, support structures, and investments, we computed expected warfighter capability levels relative to a range of deployment and transportation times. Additionally, we assessed relative system cost implications in terms of spares, transportation, and labor expenditures over a 15-year time horizon—the expected life of the program. Our analysis shows that the decision to centralize or decentralize LANTIRN repair operations hinges not on the expected system costs but on the capability and risk levels the Air Force is willing to accommodate in its operational plans. So again, we use these capability metrics in the body of this report and discuss comparative cost implications in the appendices.

ANALYSIS OF THE FUNDAMENTAL FACTOR—TIME

When weighing the implications of decentralized or centralized support, one must consider the deployment and inter/intratheater transportation times associated with each option. Forecasting this time element for MTW scenarios is difficult, so we assessed expected capability levels relative to a range of both deployment and transportation times. Figure S.1 illustrates our results for targeting pods supporting a two-coincident-MTW scenario. We show only the targeting pods because they are mission essential and generate greater demands on the maintenance system.

Given an inherent pod inventory constraint, we begin by setting a pod availability goal for both engaged and non-engaged aircraft. Availability is defined as the number of serviceable pods available for use on aircraft for specific missions. The Air Force currently does not have an availability goal for LANTIRN pods on aircraft, so we chose a value (0.8 pod per aircraft flying surge operations) somewhat higher than that used for the entire aircraft fully mission capable (FMC) rate. Using AWOS wartime removal rates and illustrative wartime flying profiles, we computed pod availability as a function of support structure performance.

Figure S.1(a) shows the expected pod availability for engaged and non-engaged aircraft (trainers) as a function of deployment time for a decentralized support structure during the second war in a two-MTW scenario. Here we define deployment time as the number of days it takes to set up functional repair operations at the FOL once surge missions begin. We indicate two decentralized support sensi-

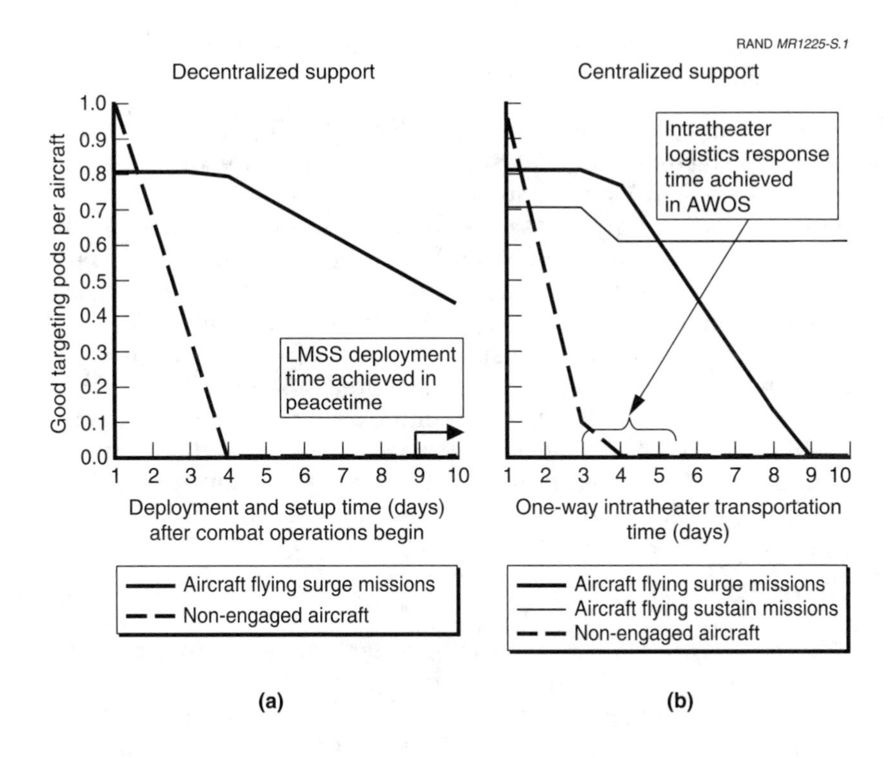

Figure S.1—Targeting Pod Availability in a Two-MTW Scenario

tivities. First, the repair system is sensitive to how quickly support equipment and capability are deployed to the theater. Second, assuming support capability deploys *prior* to the start of combat operations, if a single tester at an FOL fails, then repair capability at that particular FOL becomes very sensitive to the supply system's responsiveness in repairing that one tester. Note that if deployment takes longer than four days during the second MTW, there may be no pods available to fly training missions. Furthermore, if deployment times increase beyond this breakpoint, the Air Force risks degrading pod availability to the *engaged* aircraft. The current LMSS requires approximately 10 days to deploy and set up in theater, whereas the proposed upgrades may enable deployment and setup times under four days. Thus, deployment planners need to prioritize strategic airlift, which may be severely constrained in a two-MTW scenario, to accommodate this requirement.

The centralization option, shown in Figure S.1(b), introduces a different time factor in our analysis. Now, *transportation* time (defined as Order and Ship Time—assumes no backorders) becomes the critical system sensitivity. Because equipment and some people are prepositioned near areas of potential conflicts, deployed units must transport unserviceable pods to the regional repair operation. Again, we computed targeting pod availability during the second MTW as a function of the one-way transportation time from an FOL to a regional repair facility. Here, the critical breakpoint is four days, beyond which *engaged* aircraft capabilities may degrade. Data collected in the AWOS indicate that intratheater transportation between centralized repair operations and FOLs ranged from three to five days (Logistics Response Time [LRT], includes backorders).

STRUCTURE TRADEOFFS

Strategic and operational risks. Centralized operations may be more susceptible to terrorist attacks or may be located too far from yet unforeseen contingencies, whereas the decentralized support structure is sensitive to the availability of deployment airlift during the early phases of large-scale missions. Both structures may suffer if resupply times do not meet the performance assumptions used to set spare-parts levels. Operationally, a decentralized structure is sensitive to tester downtime. If a single set of testers is deployed, a breakdown will temporarily halt repair. In a consolidated structure, the greatest operational risk is Order and Ship Time, as discussed earlier. The severity of the effects of subpar performance depends upon how actual resupply time differs from the assumptions used to plan Readiness Spares Packages (RSPs) and pod kits for a deployment package.

Deployment footprint. Among the goals of the Expeditionary Aerospace Force (EAF) are deployment predictability to improve stability in the personal lives of Air Force personnel and quick-hitting expeditionary operations. These goals require rapid deployment of strong combat forces, with a premium on reducing footprint. While consolidation options may reduce personnel deployments by over

100 people, the greatest footprint reduction is realized through the elimination of equipment movement.

Organizational issues. Although the thrust of this analysis is on the quantitative issues associated with various logistics structures, we cannot overlook the less-tangible cross-organizational implications of our dipole options space. Decentralized support requires that individual squadron or wing commanders compete for valuable airlift early in the campaign. Competing not only with other LANTIRN units but also with other commodities, mobilization plans may need to be modified to prioritize deployment timelines. Although centralized support requires minimal tactical airlift (pods are relatively small), commanders would have to share a global asset pool that includes not only personnel and repair equipment but also tactical transport and the pods themselves.

SUPPORT OPTION ADVANTAGES AND DISADVANTAGES

Although the centralized option requires fewer test sets and fewer highly skilled personnel, the annual transportation costs may be higher. Our analysis shows that these annual costs, coupled with labor expenses, could be virtually the same across the options analyzed. The recurring peacetime costs may thus essentially be equal.

The regional support structure drastically reduces the deployment footprint, and because FSLs are removed from combat operations, both support equipment and people face lower risks. Although regional operations may become more vulnerable to attack (both conventional and cyber), proper preparations and communications design can alleviate these threats.

Collocation of test equipment not only reduces the effects of single-string failures but also eliminates the need to transport repair equipment to support various contingencies. Support equipment spares can be cannibalized from collocated sets, minimizing the effects of supply system delays. Because test set transport and setup times can be lengthy and equipment readiness is unpredictable in the theater, the regional structure offers a much more stable support system. However, daily pod transportation risks increase with the consolidated options. Pods will pass through additional transportation channels, with more people involved in the loading and unloading process. We have no data indicating pod sensitivity to transport, but rough handling in the new channels may become a problem in the proposed regional structure. Standardized training procedures and tools can mitigate this potentiality.

Finally, the consolidated intermediate repair structure will require new organizational processes. Unit commanders will have to relinquish some of their control over LANTIRN pods. They will also have to communicate closely with the support centers and other bases serviced by the same regional facility. Performance metrics and incentive systems may also need to change to support a system focused on customer (warfighter) satisfaction, on-time delivery, and quality workmanship.

CONCLUSIONS

Our analyses show that given representative planning scenarios and deployment and transportation processes, the Air Force must invest in support equipment upgrades regardless of the support structure, and it must evaluate the resource constraints it could face in a two-MTW scenario. Furthermore, the extended pipelines necessary in centralized support exclusively from continental United States (CONUS) facilities may reduce warfighter capabilities. Thus, in assessing centralized repair alternatives, the Air Force should consider networked FSL and CSL structures.

The FSL structure introduces new risks to the Air Force, but it also offers some distinct advantages over the current system. The most viable structure our analyses identified would use two FSLs and one CONUS facility. Figure S.2 shows a notional implementation of such a structure with prepositioned sets in each region (gray bubbles) and peacetime manning (white bubbles). This regional system requires that pods be shipped from FOLs to the centralized repair facilities. Again, using AWOS removal rates and an illustrative two-MTW scenario, the map indicates that the Air Force needs more test sets than it possesses. Thus, if the Air Force is to meet multiple major contingency requirements, support equipment capability should be reevaluated.

Clearly, the Air Force may want to consider alternatives not discussed in this report. Our intent is to provide an operational capa-

RAND MR1225-S.2

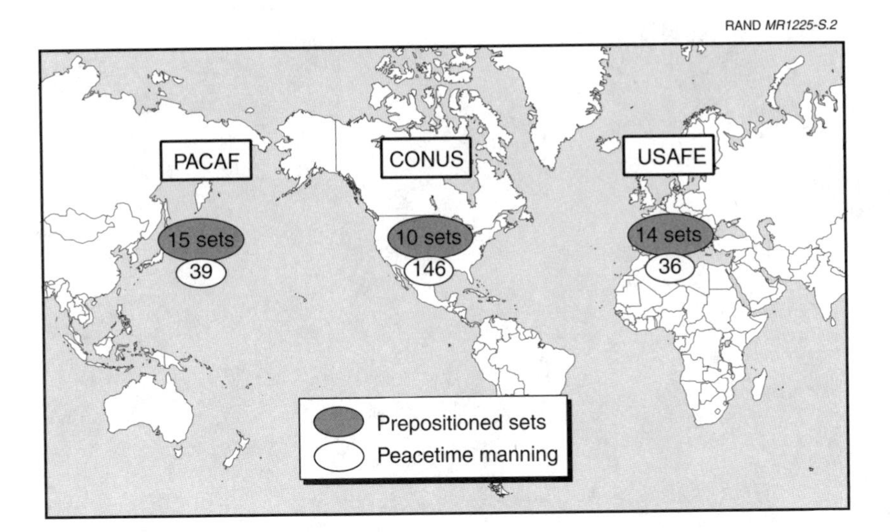

Figure S.2—Alternative Positioning of LANTIRN Testers and Personnel

bilities and risk assessment framework with which other options may be evaluated.

We recommend that the Air Force invest in the support equipment upgrades and leverage lessons learned from the AWOS, where consolidated repair operations at RAF Lakenheath, Aviano Air Base, and Spangdahlem Air Base supported deployed units, minimizing the need to transport pod intermediate-level maintenance equipment. Recall that a centralized system will be sensitive to transportation times and may suffer from poor cross-organizational cooperation and communication. Regional structures may relax some of the constraints put on the repair system, but the Air Force should be mindful of the limitations it may face in support of two MTWs.

ACKNOWLEDGMENTS

Many persons inside and outside the Air Force provided valuable assistance and support to our work. We thank Lieutenant General William P. Hallin (then AF/IL) for initiating this study and Lieutenant General John W. Handy, his successor, for continuing the sponsorship.

We have enjoyed support for our research from Air Force Major Commands responsible for implementing the Expeditionary Aerospace Force (EAF). Major General Dennis Haines (ACC/XR) and Brigadier General Terry Gabreski (USAFE/LG) provided access to personnel and data at their operating locations. Lieutenant General Cranston (AFMC/CV) and Colonel Robert Dehnert (AFMC/LG) encouraged us to show how our research could provide insights for the Air Force Materiel Command in supporting EAF implementation.

At the Air Staff, we thank Lieutenant General Michael E. Zettler (HQ USAF/IL), Mr. Grover Dunne (AF/ILM), Brigadier General Billy Stewart (AF/ILS), Brigadier General James P. Totsch (AF/ILS), Brigadier General Quentin L. Petersen (AF/ILT), and Ms. Susan O'Neal, as well as their staffs, for their support and critique of this work.

The people in the Air Force's LANTIRN maintenance shops, System Program Offices, and Air Combat Commands made this analysis possible through their sharing of data and insights. At Warner Robins Air Logistics Center, we thank Mr. George Spencer (Precision Attack Systems Program Director), Major James Grant, Master Sergeant Paul Miller, Captain Mark Laidler, Technical Sergeant Arlen Dale, Mr. Steve Kotzum, and Mr. Tom Shoemaker. At Air Combat Command, we thank Lieutenant Colonel Ray Dennie, Senior Master Sergeant

John Abdale, and Master Sergeant Joseph Eckersley for their data collection and development of operational requirements for the scenarios we modeled. Senior Master Sergeants Richard Perrier and Paul Bull as well as Technical Sergeant Michael Moffatt provided Air National Guard insights for our investment options. At the Mountain Home LANTIRN repair facility, Technical Sergeant Thomas Rowles and Sergeant Hun Schmidt helped us collect data. Captain Marc Jamison (Operations Group) and Captain Scott Miller (F-15E pilot) also supported our research at Mountain Home. We thank Chief Master Sergeant John Drew and Senior Master Sergeant Erik Mazlik at the Air Force Logistics Management Agency (AFLMA) for assisting in many of our site visits and helping us collect data used in the analysis.

Our research has been a team effort with the AFLMA, and AFLMA support has been critical to our work. We wish to thank Colonel Richard Bereit (AFLMA/CC) and Lieutenant Colonel Mark McConnell (AFLMA/LGM) for their help.

Finally, we thank Colonel Rodney Boatright (AF/ILXX) for his encouragement and support. At RAND, C. Robert Roll and Laura Baldwin have helped with reviews and critiques of our work. We also thank Gina Sandberg for her patience and promptness in preparing many drafts and Clifford Grammich for his help in writing this manuscript.

ACRONYMS

ACFT	Aircraft
ACS	Agile Combat Support
ADK	Advanced Deployment Kit
AEF	Air Expeditionary Force
AFLMA	Air Force Logistics Management Agency
AWOS	Air War Over Serbia
AWP	Awaiting parts
CSAF	Chief of Staff, United States Air Force
CND	Could-Not-Duplicate
CONUS	Continental United States
CSL	CONUS Support Location
DPG	Defense Planning Guidance
EAF	Expeditionary Aerospace Force
EOTS/BRITE	Electro-Optical Test Set/Battlefield Reconfigurable Instrument for Test of Electro-Optics
FOL	Forward Operating Location
FSL	Forward Support Location
LANTIRN	Low Altitude Navigation and Targeting Infrared for Night
LIATE	LANTIRN Intermediate Automatic Test Equipment
LMI	Logistics Management Institute
LMSS	LANTIRN Mobility Shelter Set

LRT	Logistics response time
LRU	Line replaceable unit
MC	Mission Capable
MTW	Major Theater War
NAV	Navigation
NEA	Northeast Asia
OST	Order and Ship Time
PACAF	Pacific Air Force
PGM	Precision-Guided Munition
RSP	Readiness spares package
SE	Support equipment
SRU	Shop replaceable unit
SWA	Southwest Asia
TPFDD	Time Phased Force and Deployment Data
TRG	Targeting
USAFE	United States Air Force in Europe
WMP	War Mobilization Plan

Chapter One

INTRODUCTION

Although its definition is not yet final, it is clear that the Expeditionary Aerospace Force (EAF) concept will play a central role in future United States Air Force (USAF) operations. The EAF relies on rapidly deployable, immediately employable, and highly flexible forces to serve effectively the strategic role of deterring and responding to aggressive acts. The EAF allows the Air Force to use sustainable force tailored to individual contingencies.

Earlier RAND reports (see the Preface) have discussed the importance of Agile Combat Support (ACS) in meeting EAF system requirements. Our analyses have relied on an employment-driven analytical framework—a framework identifying mission resource needs and adjusting mission goals to available resources—to guide the design and evaluation of ACS systems.[1]

This report is one in a series using this analytical framework. It focuses on the Low Altitude Navigation and Targeting Infrared for Night (LANTIRN) maintenance system, and how the system can be improved for the EAF ACS.[2] Our preliminary work looked at LANTIRN support requirements in the most stressing scenario; here we include other scenarios as well as analyses based on data

---

[1]For a detailed discussion of this work see Robert S. Tripp, Lionel A. Galway, Paul S. Killingsworth, Eric Peltz, Timothy L. Ramey, and John G. Drew, *Supporting Expeditionary Aerospace Forces: An Integrated Strategic Agile Combat Support Planning Framework*, RAND, MR-1056-AF, 1999.

[2]The LANTIRN system includes pods mounted on aircraft and their associated second-level repair resources.

collected in the Air War Over Serbia (AWOS). The new data enable a more realistic assessment of wartime support requirements.

Following our initial analysis, Major General Dennis Haines (ACC/XR) and Mr. George Spencer (Program Manager of the Precision Attack Air Logistics Center at Warner Robins) asked us to consider other variables affecting LANTIRN maintenance issues for EAF operations: the effects of performance improvements through investment, different logistics structures, multiple wartime scenarios, and the sensitivity of our analyses to assumptions on removal rates, personnel, productivity, transportation times, and equipment availability.

As the Air Force adopts EAF concepts, it faces multiple alternatives for addressing LANTIRN system degradation and obsolescence of spare parts. Alternatives include maintaining the current support system and its support equipment, modernizing LANTIRN support equipment, and developing new navigation and precision attack systems to replace LANTIRN. We examine intermediate-level pod repair and assess opportunities for consolidating some of these repair operations in regional support centers. This analysis addresses three support equipment investment options and six logistics structures in light of possible equipment performance levels and four illustrative operational scenarios that represent a broad range of possibilities that may need to be supported in the future..

EAF GOALS AND REQUIREMENTS

Several trends have led the USAF to reconsider its operational concepts. In contrast to the Cold War years in which the USAF sought to contain one major adversary at a relatively fixed number of identifiable locations, the USAF now faces much more uncertainty in its operations. A growing number of frequent, small-scale, and rapid U.S. deployments have exacerbated this uncertainty and have made clear that future operational requirements will be very different from those that led to the planning and development of the existing support system.

The current system was designed with an extensive overseas infrastructure to support one large conflict in Central Europe or Korea.

Political pressure to reduce U.S. forces permanently stationed overseas, however, coupled with economic pressure to reduce defense outlays, has resulted in basing a larger percentage of a smaller force structure in the continental United States (CONUS). This shift occurred without corresponding changes in organization or equipment. As a result, the Air Force is now straining to sustain readiness while meeting a more demanding peacetime environment of frequent deployments. This strain and continuing struggles to meet tight deployment timelines led Air Force leaders to examine alternative operational organizations and support concepts.

The growing reliance on CONUS-heavy basing coupled with the need to project force rapidly overseas present significant support challenges. The Air Force must be able to deploy aerospace capability quickly and employ that capability immediately; to meet tight deployment and employment timelines, units must be able to deploy rapidly to the reception sites and set up logistics production processes quickly. The need for rapid deployment of massive forces leads the Air Force to minimize associated support resources, particularly so that more combat forces can deploy in a given period. Demanding employment scenarios further lead the USAF to ensure that support resources are in place to sustain heavy combat operations almost immediately. At the same time, uncertainties about access to foreign bases, the resource requirements of future operations, and the difficulties in protecting forward locations favor minimizing the amount of materiel prepositioned at reception bases.

These contradictory pressures require a transportation pipeline both to reduce the support footprint (the amount of initial airlift space needed to transport operating materiel and combat equipment) and to ensure responsive resupply to support operations. This alternative would trade substantial early airlift capacity devoted to moving support equipment for constant and much smaller airlift capacity dedicated to quickly moving spare parts for the duration of the conflict.

The variety of operations that the USAF must meet presents additional support challenges. The USAF must maintain readiness for potential Major Theater Wars (MTWs) while having forces available for "boiling peacetime commitments."[3] The support system must be able to accommodate EAF operations in a variety of locations with varying infrastructure capabilities in any area of responsibility. It must be able to respond to changing events and to shift rapidly between different kinds of operations.

All these challenges have led the Air Force to reexamine its combat support system and to determine how these new support challenges can best be met.

LANTIRN SUPPORT ISSUES FOR THE EAF

Our research focuses on LANTIRN intermediate-level support systems and structures, for several reasons. LANTIRN support easily lends itself to new support structures such as consolidation that may improve the effectiveness and efficiency of the overall ACS, and thereby of the EAF. Beyond its lessons on overall support structure, this research may offer more specific insights on dealing with support issues relating to aging equipment and technology obsolescence—although they remain an essential part of combat operations, LANTIRN pods are becoming obsolete and eventually will be replaced by newer technology.

The LANTIRN system is composed of two independently operated pods mounted under the fuselage of an aircraft (Figure 1.1). The navigation pod (NAV) enables pilots to fly at low altitudes, even in limited visibility, and thus avoid detection by unfriendly forces. The targeting pod (TRG) illuminates targets for precision-guided munitions (PGMs). The TRG is key to precision attack capabilities of combat aircraft and offers distinct advantages over munitions guided by Global Positioning Systems. Laser-guided bombs using LANTIRN targeting, for example, are more effective than satellite-guided munitions against moving targets.

The USAF currently has three aircraft types configured for LANTIRN: the F-15E, F-16C, and F-16D blocks 40 and 42. Although there are several initiatives to modify additional F-16 models for LANTIRN

---

[3] "Boiling peacetime" is a term coined by General John Jumper to describe the requirements to deploy substantial aerospace forces during peacetime to ensure global stability.

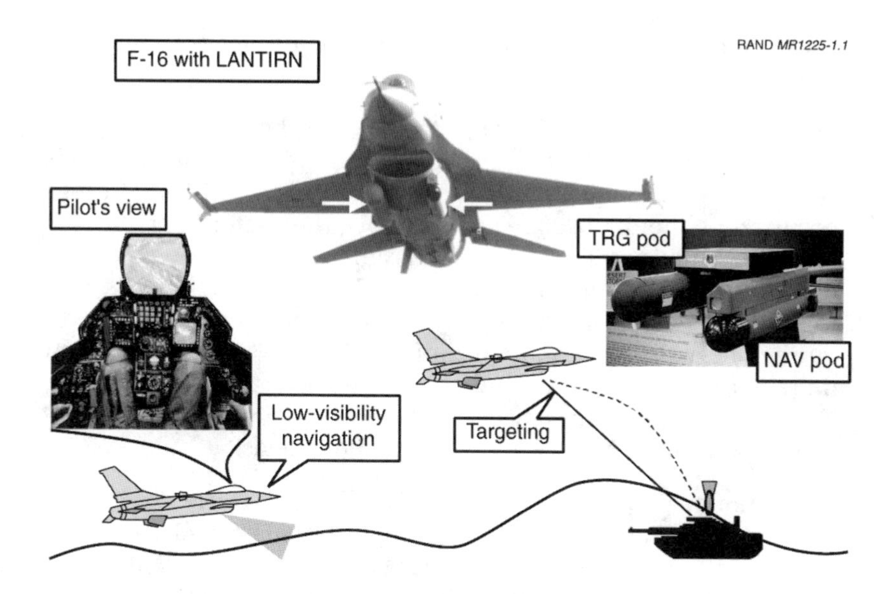

Figure 1.1—The LANTIRN System

capability, this study considers only aircraft currently configured for LANTIRN.

There are several issues affecting the future availability of this system and its support resources. LANTIRN pods and their support equipment are based on 15-year-old technology. The support technology is increasingly unreliable, and growing obsolescence of spare parts makes it increasingly difficult to repair both pods and test sets. Given the current attrition rate of five targeting pods per year, by 2002 there may be fewer pods than LANTIRN-capable aircraft in the Air Force inventory. Pods are typically lost when an aircraft crashes. Although the Air Force continues to buy F-16s and F-15Es, it is not continuing to purchase LANTIRN pods. The USAF must maintain a support system for the remaining pods to realize their maximum use.

Requirements to maintain a support system for increasingly obsolete technology are complicated by differences in availability, performance, and use of navigation and targeting pods. There are many more NAV than TRG pods. TRG pods, however, have a failure rate about twice that for NAV pods and their repair times are about four times longer.

Furthermore, our analysis of the AWOS data shows that NAV pods may not be used at all for certain operations, whereas TRG pods consistently support mission requirements. Most TRG pod usage occurred at about 10,000 feet, well above the useful range of NAV pods. NAV pods on F-15E AWOS missions flown from Aviano were used so little that by the second week of the campaign the pods were removed to reduce aircraft weight. Certain terrain and mission requirements may still dictate use of NAV pods, but it appears that there is a USAF mission trend toward reduced use of these pods. Nonetheless, our results reflect the resource requirements associated with supporting both navigation and targeting pods.

In addition to issues of LANTIRN technology obsolescence, the USAF faces increased attrition of the skilled personnel needed to support this equipment. The unique support needs of LANTIRN and the increased attrition of LANTIRN support personnel suggest that the USAF should examine all opportunities to mitigate the effects of these problems on the future readiness of the force.

DIMENSIONS OF LANTIRN SUPPORT DECISIONS

Beyond the general issues framing our research, we consider several specific variables affecting LANTIRN support decisions, as shown in Figure 1.2. First, we analyze variables affecting LANTIRN support system performance. These include likely EAF scenarios, trends in LANTIRN support personnel and equipment, and the sensitivity of LANTIRN performance to assumptions about pod employment and support processes.

The EAF scenarios we consider are

- "Boiling peacetime"—deployment of squadrons for Expeditionary Aerospace Force (EAF) peacetime commitments (such as an Air Expeditionary Force [AEF])
- "Stressing"—the resources needed for immediate employment
- "Halt phase"—the resources needed for compressed deployment to bring enemy aggression to a halt

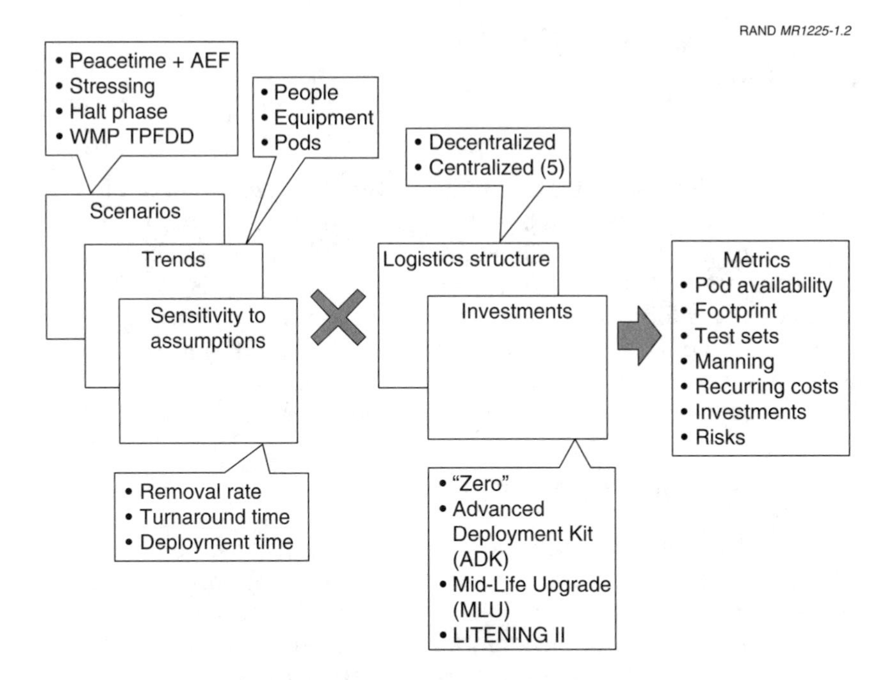

Figure 1.2—Decision Trade Space Elements for the Analysis

- An extended MTW deployment.

The trends in LANTIRN support that we focus on concern personnel, support equipment, and total number of pods. Removal rate, deployment time, and repair turnaround time (including pipeline transportation time) sensitivities are those most crucial to our analysis.

We consider both decentralized and centralized support structures, with four investment options. For each option, we analyze seven measurements for achieving EAF objectives, including metrics on

- pod availability
- deployment footprint
- test set requirements
- personnel needs

- recurring program costs
- investment costs
- risks.

These variables and operation metrics were analyzed in light of EAF goals: reducing deployment footprint, cutting operational risks to equipment and personnel, lowering peacetime operating costs and investments, and, most important for this analysis, achieving LANTIRN availability across various operational scenarios. The combination of variables on system performance, support structure, and operation metrics yield, in the end, 48 options for LANTIRN support from which we discuss the four most viable solutions.

Support Structure Goals

We evaluated resource allocation options in light of four operational scenarios, six logistics structures (five centralized options) and three investment options (LITENING II[4] investment does not affect LANTIRN resource requirements). Specifically, we analyzed the distribution of intermediate test stands, the associated personnel, and spares in terms of pods and line replaceable units (LRUs). We also assessed the peacetime transportation system costs needed to support various logistics support structures. These parameters were analyzed in connection with EAF objectives, particularly achieving LANTIRN availability across various operational scenarios. Other objectives include a reduced footprint, reduced operational risk to both the support equipment and personnel, and lower peacetime operating costs and investments.

Examining the spectrum of operational requirements produces a set of goals for designing the support structure. We start with the two-MTW mission from the DPG. Current Air Force thought is concentrating on meeting a potential MTW with massive, immediate force in order to bring about the "halt phase." Such force is typically considered to require a suite of high-precision weapons now avail-

[4]A targeting pod similar to LANTIRN but supported with a two-level maintenance system.

able in only limited quantities or still in development.[5] As one might expect, such weapons are extremely expensive. Budget cuts over the last few years coupled with an environment that makes budget increases politically difficult force the Air Force to make tough budget choices. The Air Force and the other services are looking for ways to reduce costs in order to finance important programs ranging from acquisition to personnel retention. Hence, cost minimization becomes a goal in redesigning the support structure.

However, cost minimization must be secondary to maintaining or increasing operational effectiveness. Our employment-driven models address the first part of this task—determining the minimum resource levels to meet operational demands. However, the manner in which the resources are composed can have an effect on operational effectiveness.

As previously discussed, the fewer support resources deployed in the early stages of a halt-phase operation or an AEF, the more capacity is available for moving valuable combat forces to the theater. Minimizing the support structure's deployment footprint thus becomes another goal. Because of the new operational concepts, all missions have the potential to require tight timelines. A third goal, then, is to meet the required response time. The Chief of Staff of the Air Force (CSAF) has set this goal at 48 hours.

Each alternative support structure is designed to support anticipated operating demands, but the levels and types of operational risks vary across the alternatives. For each structure, we must identify the major risks and evaluate their probabilities and effects.

Finally, the rigorous cycle of deployments for boiling-peacetime commitments (such as the Air War Over Serbia and Operation Northern Watch Over Iraq) has increased personnel turbulence, making it difficult to retain skilled aircraft technicians. Reducing personnel turbulence by regular scheduling of deployments or balancing deployment requirements among units is a central goal of the

---

[5]See David Ochmanek, Edward Harshberger, and David Thaler, *To Find, and Not to Yield: How Advances in Information and Firepower Can Transform Theater Warfare,* RAND, MR-958-AF, 1998, for a description of the halt phase problem.

EAF. We adopt further reductions of personnel turbulence as a goal of LANTIRN support structure design.

Support Structure Locations

We compared the operational performance and resource requirements of the current LANTIRN support structure with those that could be attained by five different regional structures composed of forward support locations (FSLs). In the current structure, repair capability deploys with the units. In the regional options, FSLs and one or more CONUS locations (CSLs) provide repair. During deployments, the CONUS facility would support only nondeployed aircraft flying at peacetime sortie rates.

The spare parts requirement for the current structure depends on how quickly repair operations can be established in the theater once the support equipment is deployed. For the regional options, this requirement depends on the transportation time between FSLs and forward operating locations (FOLs). The longer it takes to ship and return items for repair between FOLs and FSLs, the greater the spare parts requirement for centralized structures.

Our calculations assume that the Air Force does not procure additional pods. As stated earlier, there will be fewer targeting pods to support the future scenarios we analyzed. We assessed potential policy options to alleviate this apparent shortfall. One approach is to reduce the number of pods available per Primary Aircraft Assigned (PAA) while maintaining the same LANTIRN flying program. For every 100 aircraft, for example, there might be only 80 combat-capable pods available at the end of a flying period. Some LANTIRN pods may need to be moved from one aircraft to another to achieve the desired LANTIRN missions with fewer pods. Conversely, we argue that at any given point some aircraft will not be mission capable because of other factors, thus reducing the actual pod requirement for the unit. We employed this availability methodology in our computations.

Consolidated and decentralized structures pose different types of risks. Under the current, decentralized concept, most FOLs likely will have single test stations. If a single test set goes down and cannot be repaired quickly, then the planned maintenance resupply will be delayed. Decentralized structures also face risks posed by delays in deployment and in-theater setup time. Under consolidated structures, the collocation of several test stations nearly eliminates the risk of having no maintenance resupply. Consolidated repair also, however, relies on resupply through inventory and transportation rather than local maintenance. Consolidation requires close management of the distribution system and shared assets such as pods or LRUs.

Consolidating maintenance would create a more functionally oriented organization, in which units would have to rely on others for their resources and not maintain themselves all the resources they need for operations. Although functionally oriented organizations offer many advantages, such organizations can change subunit objectives and introduce cross-functional communication problems that impede planning. Some of these issues became readily apparent during the AWOS, and so we recommend that the Air Force closely examine the logistics and organizational issues associated with centralized support.

OUTLINE

In Chapter Two, we discuss the application of our employment-driven modeling approach to LANTIRN support issues and the options for meeting them. We give an overview of the scenarios that the USAF may face, as well as the investment options it is considering to ensure sufficient LANTIRN resources. In Chapter Three, we analyze several metrics for measuring LANTIRN support system requirements, and what they indicate about the available options for future support. In Chapter Four, we review the financial implications of each of the possible LANTIRN support systems by level of consolidation and new equipment investment. In Chapter Five, we evaluate the risks and advantages of each support option and compare them across the decision trade space.

Chapter Two

ANALYTIC APPROACH, LANTIRN SCENARIOS AND OPTIONS

Our research on EAF support options uses an employment-driven analysis framework that identifies mission resource needs and adjusts mission goals and available resources to match each other. This approach is shown in Figure 2.1. The first step is to identify mission requirements or the force packages necessary to accomplish anticipated missions (i.e., the types and numbers of aircraft, weapons, and sortie rates needed). In this case, the information is used to estimate the demand for LANTIRN support capabilities, including the equipment, personnel, and other resources such as spare parts and transportation needed to provide these capabilities. These processes are shown in the left and middle portions of Figure 2.1. We then determine the costs of each alternative and evaluate whether they meet operational requirements.

Among variables we consider are recurring costs, deployment footprint, risks, and flexibility, as shown in the right portion of the figure. If the alternatives do not meet operational needs, then this framework can be used to revise operational objectives or to develop alternative support practices or technologies to overcome constraints. Supply system issues are not addressed in this study for two major reasons. First, we had no data to indicate which test equipment components drive particular mission-capable rate degradation. Second, because mission-capable rates were reported as a monthly average, it was difficult to ascertain how long the equipment was down as a result of lack of parts or other maintenance resources. Hence, we chose to model support equipment availability across a range of possible performance levels, as discussed later. The discussion ac-

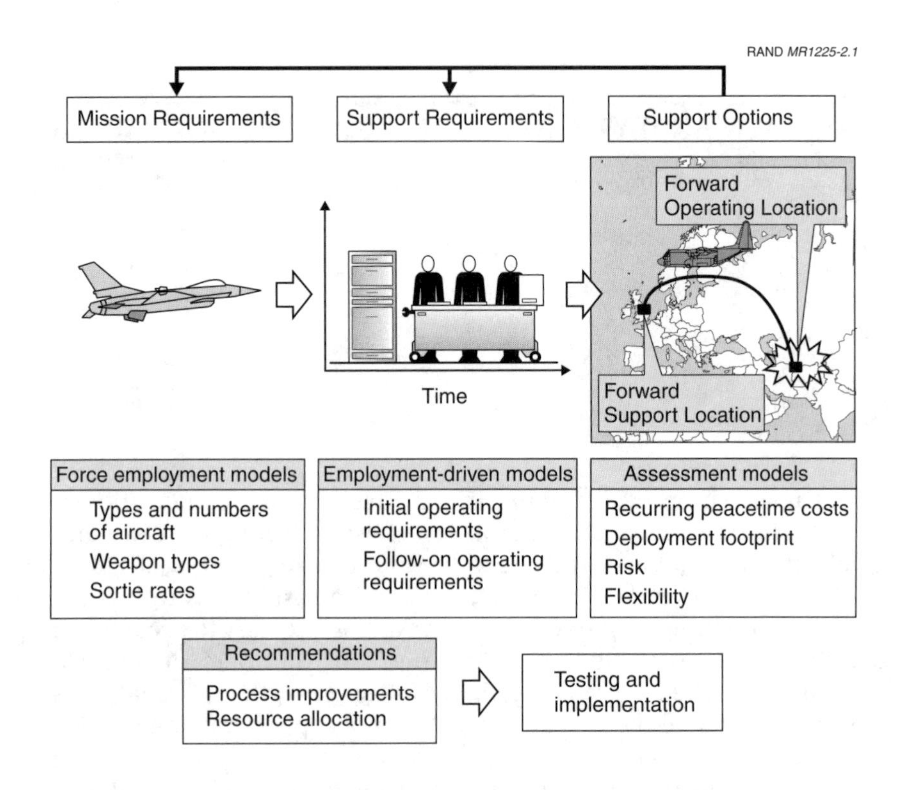

Figure 2.1—Our Research Uses Employment-Driven Models

companying Figure 3.4 in Chapter Three offers insights into the risks associated with having a single string fail at an FOL.

The alternative support structure designs are defined by peacetime and wartime locations of LANTIRN aircraft intermediate maintenance assets. These locations drive the quantities of four resources: intermediate test stands and fixtures, personnel, spare parts, and transportation assets. We extend this approach to assess multiple investment options and their effects on support equipment performance and hence resource requirements.

The LANTIRN analysis begins with employment-driven resource models to determine the minimum resource levels that enable each support structure to meet the selected scenarios. After determining the resources and composition of each structure, we evaluate that structure against both peacetime and EAF operational goals. An example of these computations is given in Appendix A.

ELEMENTS OF LANTIRN ANALYSIS

Figure 2.2 shows the basic elements of our analysis model as applied to LANTIRN employment and support structures. We used computer models (described in Appendix A) to assess the requirements for test sets, personnel, and inventory. The loop on the left side of the chart describes the system demand. Given a specific employment program, we can predict LANTIRN maintenance needs in terms of the number of pods removed from the aircraft at the flight line for back-shop repair. We modeled removals to the back shop, not those to the repair shop at the flight line. Once removed, the pods must be transported to the back shop, which may be on base or off-site. In the shop, we modeled that the pods await repair (Queue)

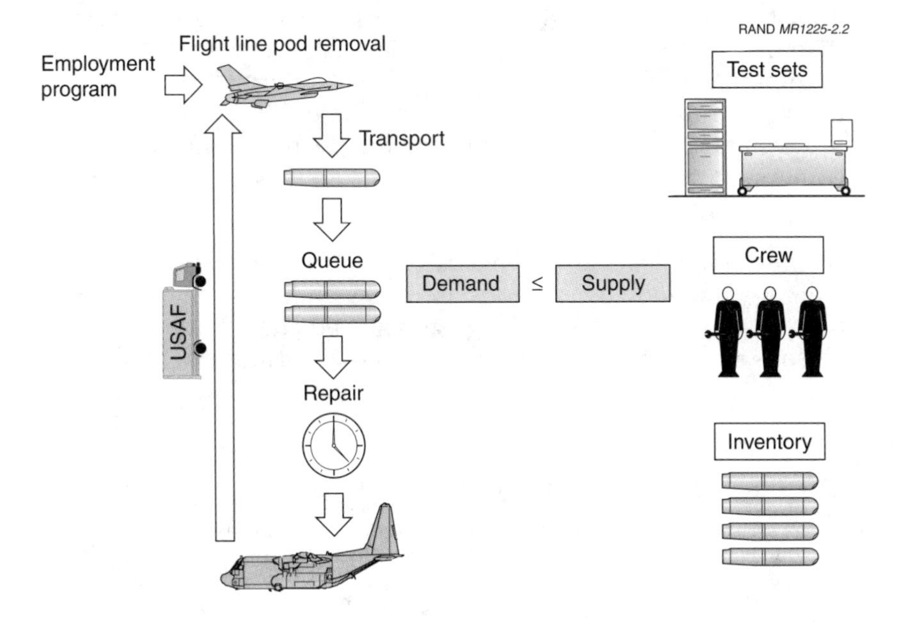

Figure 2.2—Supply and Demand Models for Assessing LANTIRN Requirements

for no more than 20 hours[1] (our imposed constraint). This value was based on inputs from the Air Force working group supporting this study because there were no data available to quantify actual in-shop queue times. The in-shop repair times that we model are based on data from Mountain Home, Elmendorf, and Moody air bases and include Bench Check Serviceable times. They are based on the Elapsed Time Indicator clock times and include powered-off repair. After repair, the pods must be transported back to the flight line (Transport in the figure), by trailer for on-base repair, or by air or truck for consolidated repair. All these processes generate demands on the system that we can measure in terms of time.

The supply side of the model, shown on the right side of Figure 2.2, depicts three major elements of supply: test sets, personnel, and inventory. Wherever possible, these are measured in terms of time (e.g., hours that personnel and test sets are available). Personnel availability depends on work schedules, productivity rates, and logistics structures. Inventory is the number of pods or LRUs that are available. In each scenario, the goal was to always have demand less than or equal to supply. We determined stockage needs by using maximum daily removal rate estimates, including lead-time variability in computing safety stock levels. We determined the number of test sets needed by using nonlinear regression methods that link equipment availability to work schedules and test set locations (an example is given in Appendix A).[2]

We determined personnel needs through industrial engineering manning methodologies. Pods are heavy and must be handled by two persons. Also, laser safety regulations dictate that at least two persons be present at a test station for repair. Thus, the base direct labor requirement is two persons for every pod or fraction of a pod

---

[1]We base this value on multiple shop visits and interviews with both shop personnel and the Air Force working group that supported this study (including five senior maintainers). Twenty hours is the typical wait time during peacetime operations. We evaluate queue length as an outcome of our models instead of a global system variable that should be optimized. Our goal in this analysis was to assess resource requirements to ensure a certain queue length rather than the number of resources required to minimize the queue.

[2]See Eric Peltz, Hyman L. Shulman, Robert S. Tripp, Timothy Ramey, Randy King, and CMSgt John G. Drew, *Supporting Expeditionary Aerospace Forces: An Analysis of F-15 Avionics Options*, RAND, MR-1174-AF, 2000.

that arrives at a shop. We augment this number based on predicted productivity rates. Military personnel typically perform many more tasks, such as training, briefings, and equipment repair, than those tasks that pertain to their formal assignment. Air Force policy documents indicate that military personnel typically have productivity levels of about 60 percent during peacetime and 90 percent during wartime.[3] All told, these manning requirements and productivity levels mean that, during peacetime, a shop requires three (direct labor) persons per shift for every pod expected to arrive on a given day.

In addition to these direct manning requirements, we examine indirect labor needs in calculating total labor requirements. Indirect labor includes trainees, supply people, shift supervisors, and shop chiefs. Peacetime operations at shops with two eight-hour shifts must have one supervisor per shift and one shop chief per shop. During wartime, shops have two 12-hour shifts. Trainees and supply personnel requirements range from one-half person to two persons per test set, depending on the support equipment used and the employment scenario. Upgraded equipment was assumed to require fewer trainees and supply personnel during wartime, based on inputs from the Air Force working group and the support equipment supplier.

We also considered trends in attrition of skilled personnel. Personnel attrition will have many effects on the LANTIRN support system. If personnel skill levels decrease, fault isolation time—the time needed to isolate and identify a pod problem—will increase. This in turn will increase the number of persons required to support a given demand level. Eventually, this may require additional test sets to support the same demand level. Modeling this effect is quite complex, and predicting future skill levels is even more difficult, so we assessed manning requirements simply by using the range of expected productivity levels in peace and war. The potential effect of reduced skill levels is discussed in greater detail in Appendix B.

---

[3]AFI38-201, "Determining Manpower Requirements," http://afpubs.hq.af.mil/pubfiles/af/38/afi38-201/afi38-201.pdf, 1999.

SCENARIOS FOR ANALYSIS

We consider an illustrative range of scenarios from peacetime operations with two deployed AEF units to two coincident MTWs to examine the robustness of LANTIRN support options. We use these scenarios to determine the costs and operational benefits of LANTIRN maintenance structures that satisfy operational requirements ranging from those posed by two coincident MTWs, one MTW, and small-scale AEF deployments in boiling peacetime operations. We found that resources satisfying a two-MTW "stressing" scenario (and the halt-phase scenario), or the immediate buildup and massive employment of forces after the start of a major theater war, will satisfy the demands of other less-demanding missions. We therefore designed the alternative structures and their resources to meet the requirements of missions for the stressing scenarios as an upper bound.

The most stressing scenario that we developed involved modeling two coincident major theater wars. In this scenario, the aircraft in the first MTW, e.g., surge for a few days and then are still flying at sustain rates, when the second MTW, e.g., Northeast Asia (NEA), would begin with aircraft flying at surge rates. Figure 2.3 shows how LANTIRN-capable aircraft would deploy for coincident MTW scenarios in SWA and NEA.

We consider three elements of each scenario. The first is the number of LANTIRN-capable aircraft deployed to each theater. The second is the type of aircraft deployed—in this case either F-15E or F-16 block 40/42. The third is the day on which each squadron or wing begins surge operations.

We do not model LANTIRN resources for all aircraft depicted in Figure 2.3. Figure 2.3 shows 144 F-16s to be employed in a second MTW in NEA, but we consider the resources needed for only 104 of these. The reason for this is that we assumed that the total number of F-16s using PGMs in two MTWs is greater than the number of LANTIRN F-16s currently in USAF inventory. The 144 F-16s that we show in the second MTW are therefore augmented in our model by 40 Air National Guard F-16s with LITENING II capabilities. All other aircraft shown in Figure 2.3 play a part in our model for resource computations.

RAND MR1225-2.3

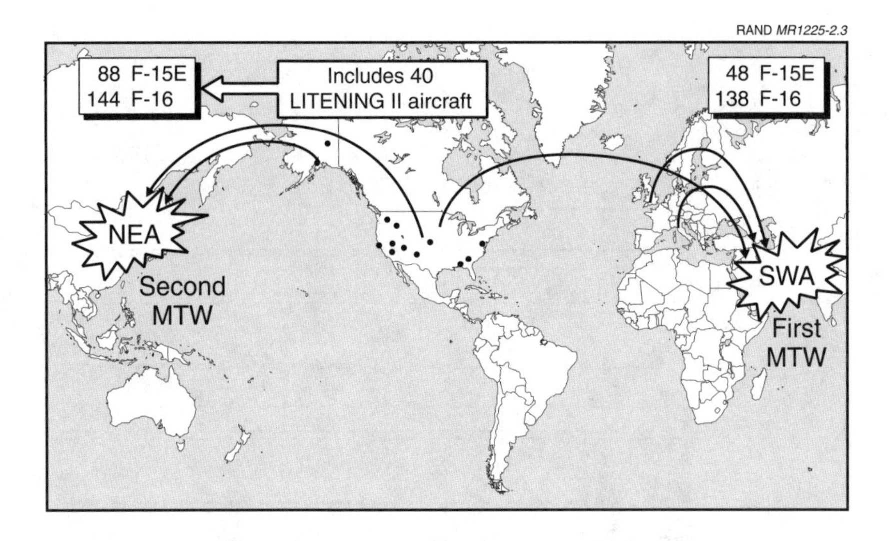

Figure 2.3—Deployment of LANTIRN Aircraft in a Two-MTW Scenario

Each scenario we consider has its own planned deployment and sortie-flying program, as depicted in Figure 2.4 for four scenarios: peacetime with AEF deployment, the two-MTW stressing and halt scenarios, and extended deployments. Each flying program has three similar elements. At the left-hand side of each graph, all aircraft start by flying peacetime sortie rates. A contingency, which can range from a boiling-peacetime operation to an MTW, requires aircraft deployment followed by a programmed employment profile. We use illustrative employment profiles for all aircraft that engage in combat operations consisting of a surge period followed by sustained operations. While some aircraft are engaged in one region (shown by a black solid line and dashed line), other aircraft remain at peacetime sortie rates (shown by a dotted line). As LANTIRN-capable aircraft deploy to regions one and two, the number of non-engaged aircraft drops—thus maintaining a constant total global inventory of LANTIRN-capable aircraft. All of the scenarios we examined have aircraft responding at surge rates to a contingency in a second region while aircraft in the first region fly at sustain rates. Finally, aircraft in both regions fly at sustain rates, with aircraft not in these regions (typically in CONUS) continuing at peacetime rates.

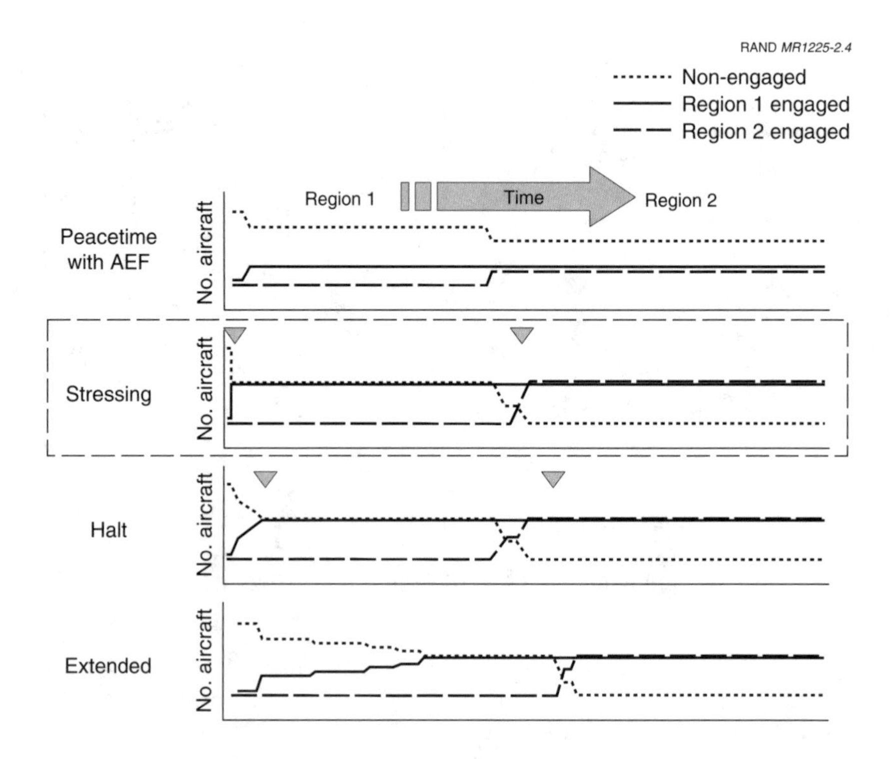

Figure 2.4—Sortie Flying Programs for Deployed Aircraft

The number of aircraft deployed to each region, as well as the timing of their program for peacetime, surge, or sustain rates, varies by the scenario considered. The extended scenario may be more representative of current planning assumptions, whereas the "halt" and "stressing" scenarios were designed to analyze the effects of greater strains on the support system. In every scenario we included non-engaged aircraft to meet unit training requirements.

Our calculations for resource requirements focus on the peak sortie-generation period for each scenario within each region. The arrowheads shown on the stressing and halt scenarios indicate approximately when the most resources are needed. Our analysis found that the resource requirements for the stressing and halt scenarios are identical; they differ only by when they occur. We therefore show results for only three scenarios—peacetime with AEF deployment,

MTW halt phase, and MTW extended—when describing differences in resources needed by scenario. Furthermore, we focus primarily on only two scenarios—peacetime with AEF and the MTW halt phase.

In addition to our scenario assumptions, we have several other assumptions for our analysis. We base our models on likely scenarios in 2008 to allow time to modify the LANTIRN support structure, as necessary. We model MTW needs for wars occurring first in Southwest Asia and then in Northeast Asia. We model both the effects of no technological improvements to the pods and support equipment as well as those of several equipment-upgrade options. Although our initial data collection showed no indications that pods are failing more frequently, we assessed the sensitivities of our analysis to increased pod failure rates through a simulation based on wartime removal rates computed from the AWOS data. We also assumed no wartime attrition of aircraft, which would place the most stressing demand on the support system.

Finally, we modeled decelerated removal rates in wartime—rates reflecting lower predicted levels of pod removal per sortie. To model the wartime decelerated removal rates, we employed methodologies similar to those used in calculating readiness spares package (RSP) requirements for avionics. These are based on a study by the Logistics Management Institute (LMI) that implies pod failures depend more on the number of sorties than on the sortie length.[4] Our assessment of AWOS data from Aviano and Lakenheath air bases indicates significantly higher pod removal rates during wartime flying conditions (see Appendix F). Thus, support equipment resources may need to be based on removal rates other than those developed from the LMI document. We highlight these implications here to show the risks the Air Force faces if it were to attempt supporting two coincident MTWs.

---

[4]F. Michael Slay and Craig C. Sherbrooke, *Predicting Wartime Demand for Aircraft Spares*, Logistics Management Institute, McLean, Virginia, AF501MR2, 1997.

INVESTMENT OPTIONS FOR FUTURE LANTIRN SUPPORT

Because we found no data showing pod performance to be declining, we focus on possible decreases in support equipment performance. We modeled three investment options that the Precision Attack System Program Office (SPO) is considering for support equipment upgrades. These improvements are intended to prolong the operability of the current test sets through replacement of obsolete components and subsystems. They are also designed to make LANTIRN support equipment more readily deployable.

First, we considered the case of no investment, for which we modeled support equipment (SE) availability or mission-capable rate (MC) based on projections of test set degradation over time. With no equipment upgrades, we estimated the single-string MC rate at three possible levels: 90 percent (for today's single-string MC rate) and 80 and 70 percent to assess the implication of a 10 percent incremental degradation over the next 10 years. The last two MC rate values are not forecasted numbers but a range of possible outcomes to help assess support equipment MC rate sensitivity and its effect on our results (see Appendices B and F).

Second, we analyzed the Advanced Deployment Kit (ADK) investment option. This option would entail a modular upgrade to the existing LANTIRN Mobility Shelter Set (LMSS), which would also improve pod-level repair capabilities. We calculated that pod repair time may be reduced by at most 25 percent with the ADK investment, using equipment supplier specifications and prototype test data collected from the Tulsa Air National Guard. The ADK would also improve the reliability and deployability of the support equipment. With the ADK investment, we estimate MC rates could range between 100 and 60 percent. The deployment footprint, measured in pallets, would be reduced by more than 50 percent. Most important, deployment and in-theater setup time may be reduced from more than 10 days to less than three days (assuming there is no wait time for strategic airlift). The ADK upgrade is estimated to cost about $2M per set.

Third, we analyzed the effects a Mid-Life Upgrade (MLU) option would have when used with the ADK investment to improve LRU test capability and overall support equipment performance. Again, because forecasting performance is extremely difficult given the data available, we chose a possible MC rate range (like that for the ADK option) to assess output sensitivities. Furthermore, although we found that recurring cost differences are negligible, our results show the major differences between investment (ADK or MLU) and no investment. In both the ADK and MLU options, deployed repair capability is limited to pod-level work. LRU capabilities currently within the LMSS do not deploy with the units in these options.

Figure 2.5 shows the physical configuration of the current system and the proposed upgrades. The LMSS used to test and repair pods and some LRUs is shown on the left. It is a completely self-contained system and fairly large, requiring some 5000 square feet of operating space, as the human figure indicates.

The boxes in the middle of the upper right-hand image, taken inside an LMSS, are a prototype of the ADK upgraded electronic equipment. This upgrade would be retrofit into existing slots in the LMSS (shown in gray) but could be quickly removed for deployments. The electro-

RAND *MR1225-2.5*

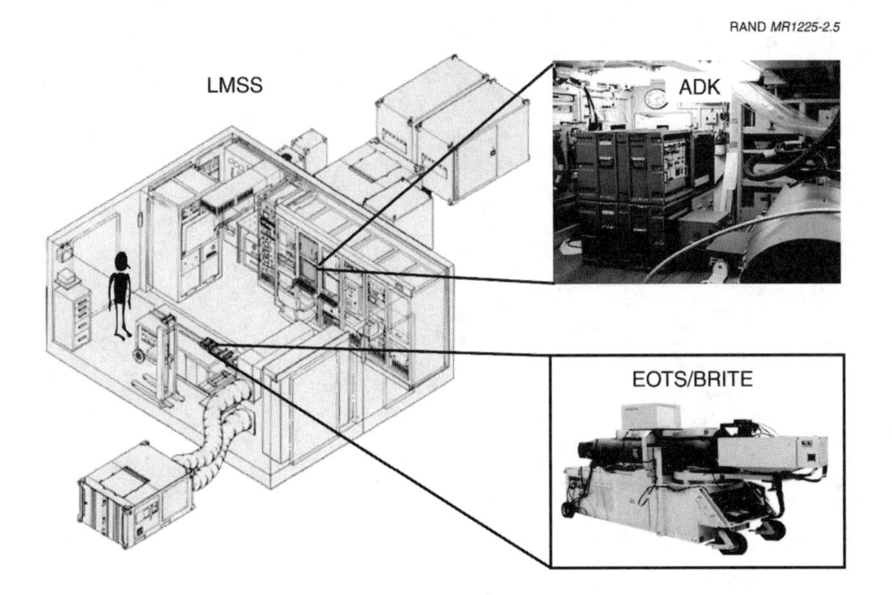

Figure 2.5—Current LMSS and Proposed ADK and EOTS/BRITE Upgrades

optical test stand (EOTS/BRITE) system shown below the ADK is an integral part of the ADK upgrade and operates with the electronic equipment described. This subsystem, which replaces an existing system within the LMSS, would be part of the deployment package. In addition to all of the elements of the ADK upgrade, the MLU would replace many of the obsolete systems used to repair LRUs in the LMSS.

Chapter Three

SUPPORT OPTIONS AND PERFORMANCE MEASUREMENTS

Conceptually, as Figure 3.1 shows, there may be three ways to support deployed units. First, repair capability can deploy with the unit, putting the maintenance shop at the FOL. Second, repair operations can be permanently located in regions where major contingencies are most expected, at FSLs. Third, repair capabilities can be kept in CSLs, and LANTIRN pods and LRUs transported to and from the FOL.

How well these concepts provide future global combat support depends upon strategic combat support design decisions about

RAND MR1225-3.1

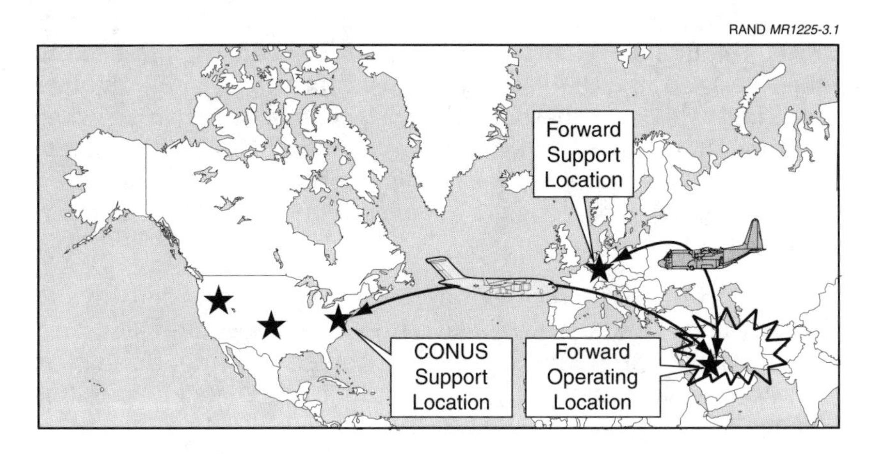

Figure 3.1—Conceptual Support Options for Deployed Units

- the number and location of FOLs with prepositioned materiel
- the resource levels for each such FOL
- the number and location of FSLs
- the functions that FSLs perform
- the base structure
- the organizational structure, and
- the transportation infrastructure.

Each combination of strategic options, or each combination of FOLs and FSLs, has different effects on operations, costs, and risks. USAF support operations traditionally have been provided through decentralized services. Decades of experience have given the USAF a full perspective on the advantages and disadvantages of such a system, so we examine how centralization of LANTIRN support compares with the current structure. Centralized or regional support may offer several advantages over the current decentralized system. Collocation of test equipment can enhance repair capacity through economies of scale and through cross-cannibalization of the support equipment (Appendix C examines test set supply and demand relationships). By having regional centers with prepositioned support equipment and established transportation routes near potential trouble spots, the Air Force can reduce deployment needs during the first stages of a contingency. By reducing the need to deploy test sets, the USAF frees up initial contingency airlift for other assets and avoids the question of tester functionality that arises when equipment is moved. Regionalization also reduces the system requirement for spare LRUs through safety stock consolidation and, by aggregating demand, reduces requirements for support personnel.

Although consolidation offers many advantages, it is sensitive to transportation delays. These delays drive pipeline requirements and, if not managed properly, can severely hinder system performance. We address transportation sensitivities in greater detail later in this chapter.

CONSOLIDATED SUPPORT OPTION STRUCTURES

The support structure options we examine range from completely decentralized to completely centralized. The current Air Force LANTIRN maintenance structure is completely decentralized, and we treat it as the base case. Each combat-coded squadron is assigned one set of intermediate maintenance assets that follows it to any FOL. Complete centralization, the opposite extreme, would consolidate all intermediate maintenance at a single CONUS location to support all peacetime and potential combat missions. Alternatives between these extremes would use different numbers of FSLs.

In the consolidated options, repair occurs at either a single CSL or at permanent FSLs able to support both combat and peacetime missions—thus the planned capacity of FSLs is based upon MTW demands. A CSL operating in conjunction with one or more FSLs must meet the peacetime demands of all CONUS-based aircraft, which means that the CSL will have excess capacity when some CONUS units are deployed overseas during war. To improve wartime maintenance efficiency, excess CSL personnel could be shifted to FSLs during contingencies. To consolidate and preposition equipment at FSLs, only technicians need to move from the CSL for wartime deployment. The equipment at the FSLs would be kept "warm" supporting boiling-peacetime operations.

Figure 3.2 shows the consolidated system structures that we analyzed. We first considered, as shown at the top, a single CONUS location with either pod or LRU repair capabilities. In this structure, pods to be repaired would be shipped from all global locations to the CSL. The Precision Attack SPO asked us to consider a variation of this structure in which a single CSL would process only LRUs, to be swapped at an intermediate shop on base and shipped to a CSL for repair. Each base or FOL would have its own ADK with limited repair capability. This option adds an echelon to the support system and does not offer significant performance improvement.

We next assessed a two-CSL structure for pods only. This system is identical in concept to the single-CONUS pod repair option but may offer some strategic advantages in supporting multiple contingen-

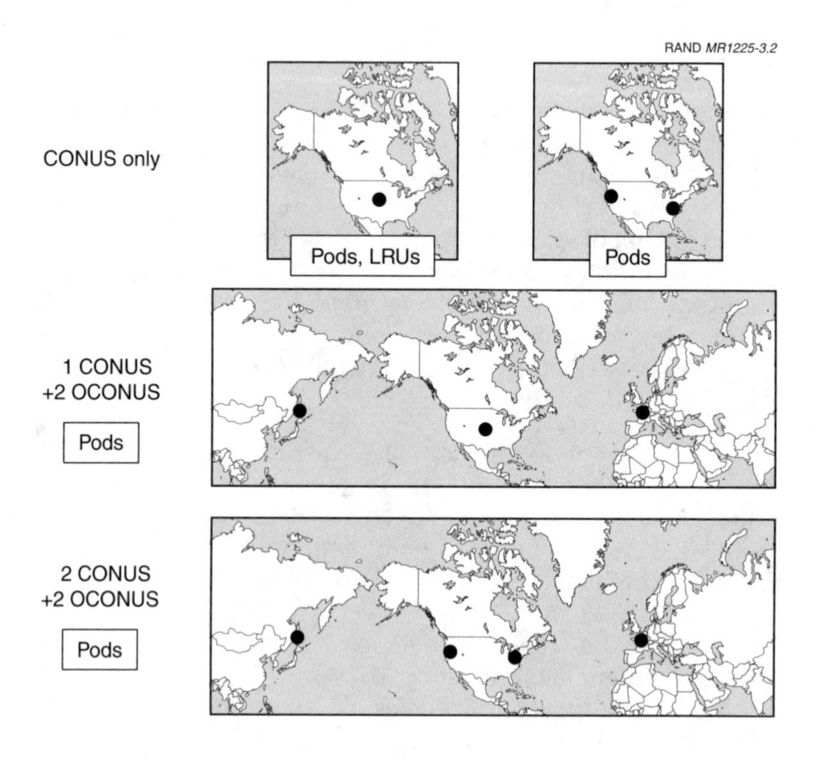

Figure 3.2—Consolidated Support Structures

cies. We also considered, as shown in the middle of Figure 3.2, a structure comprising a single CSL with two out-of-CONUS (OCONUS) locations. These FSLs, in United States Air Forces in Europe (USAFE) and Pacific Air Forces (PACAF), were modeled to support contingencies in SWA and NEA. We also assessed a structure of two CONUS and two OCONUS facilities, as shown on the bottom of Figure 3.2.

Finally, we analyzed a mixed-virtual alternative, portrayed in Appendix D. This alternative features two CSLs but with resources for three locations, one in CONUS and two OCONUS. The resources for a three-location structure are assigned to only two locations, with excess capacity deployed to other locations when a contingency is imminent. This option offers greater flexibility and several advantages of regional support, but it may be costlier and pose risks associated with equipment deployment. Again, we highlight results for the two major logistics structure options (decentralized or centralized) in the body of this report and discuss other options in Appendix F.

SUPPORT STRUCTURES AND MATERIEL MOVEMENT

The current decentralized structure has three echelons, as shown at the top of Figure 3.3. Pods are repaired at the O-level (flight line) and I (intermediate)-level operations on base or at an FOL. Broken LRUs and Shop Replaceable Units (SRUs) are shipped to an air logistics center (ALC) or depot. Replacements for the nonfunctioning subsystems are shipped from the depot to the operating base.

A centralized structure for pod repair also has three echelons, as shown in the middle of Figure 3.3. First, O-level repair is performed at the base or FOL. Second, complete pods needing I-level repair are shipped from the operating location to an FSL or CSL. Third, LRUs and SRUs are shipped from the regional center to the depot. Repaired LRUs and SRUs are shipped to the regional center from the depot, while repaired pods are shipped from the central facility to the individual units.

We also considered a four-echelon system in which each unit is equipped with limited repair capabilities. In this structure, O- and I-level technicians with deployed units perform only LRU remove and replace (R/R) operations in support of pod repair. LRUs may be swapped out of failed pods, but the individual LRUs would not be repaired on base. Nonfunctioning LRUs would be shipped to a single LRU repair facility. Only SRUs would be transported from the LRU regional repair facility to the depot. This logistics structure adds some deployment flexibility, but it increases the number of echelons in the system and does not yield significant cost savings. Again, we focus on the dipole options of centralization versus decentralization in the body of this report.

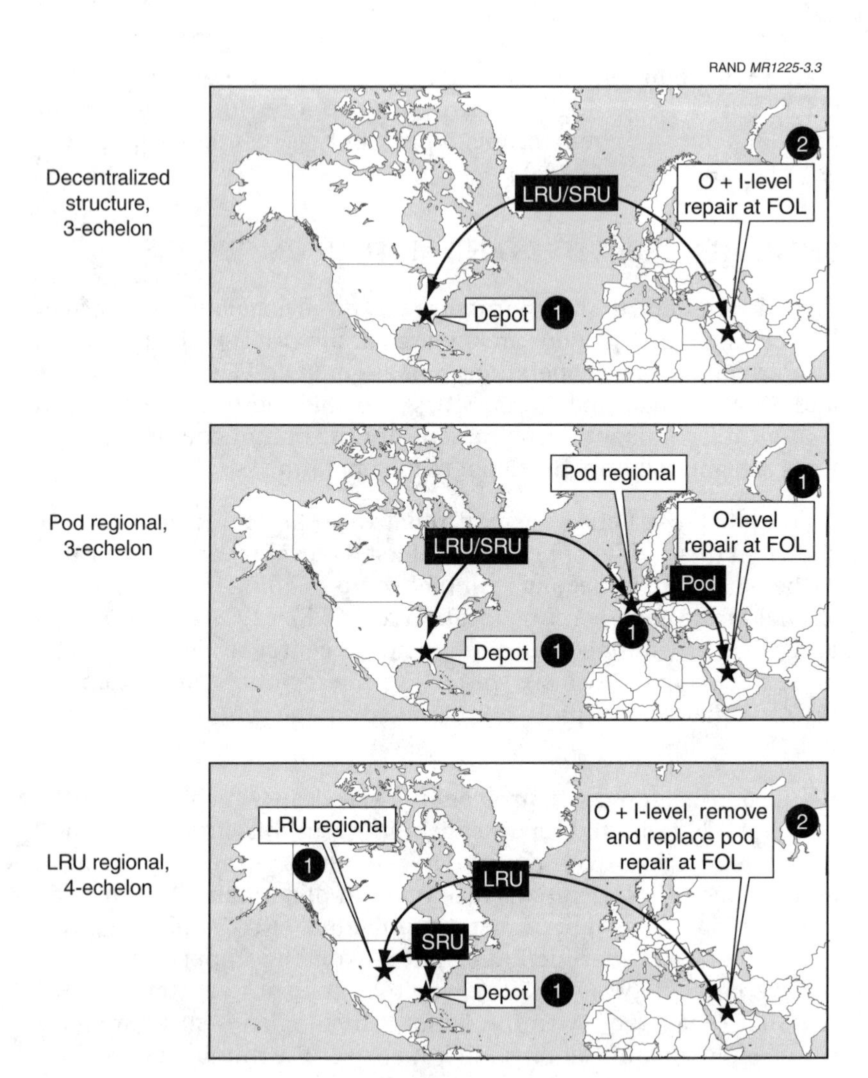

Figure 3.3—LRU Repair Consolidation Adds Another Echelon

MEASURING SUPPORT STRUCTURE PERFORMANCE: SELECTING AND ANALYZING METRICS

How can we evaluate the potential performance of alternative support structures? What metrics should we select? How might actual performance vary by slight changes in our modeling and assumptions? Below we propose metrics for seven variables to measure support system performance: pod availability, deployment footprint, test set requirements and availability, personnel requirements, recurring costs, investment costs, and operational risks. We analyze sensitivities for those variables for which only slight variation causes great effects in expected system performance.

Pod Availability

Most performance sensitivities stem from time requirements. The decentralized logistics structure is most sensitive to deployment and setup time. Regional structures are sensitive to resupply transportation times. Delays in deployment and setup for decentralized structures, or in resupply transportation for centralized structures, cut the number of pods available for combat aircraft. Requirements for RSP and pipeline and safety stocks further raise the sensitivity of each system to time delays and may lower the number of pods available for use by the warfighter. These sensitivities are most acute for TRG pods, which are used more often, fail more frequently, and take longer to repair than NAV pods.

Decentralized Structures

Figure 3.4 shows how deployment and setup times affect TRG pod availability in a decentralized support structure. Our calculations are based on removal rates computed from the AWOS data and representative flying profiles. In other words, we assumed aircraft fly according to the planned programs and experience wartime removal rates similar to those experienced in the AWOS. Under decentralized support, pods are not shared between units. Each unit must have sufficient stock to support surge operations, the most stressing flying period.

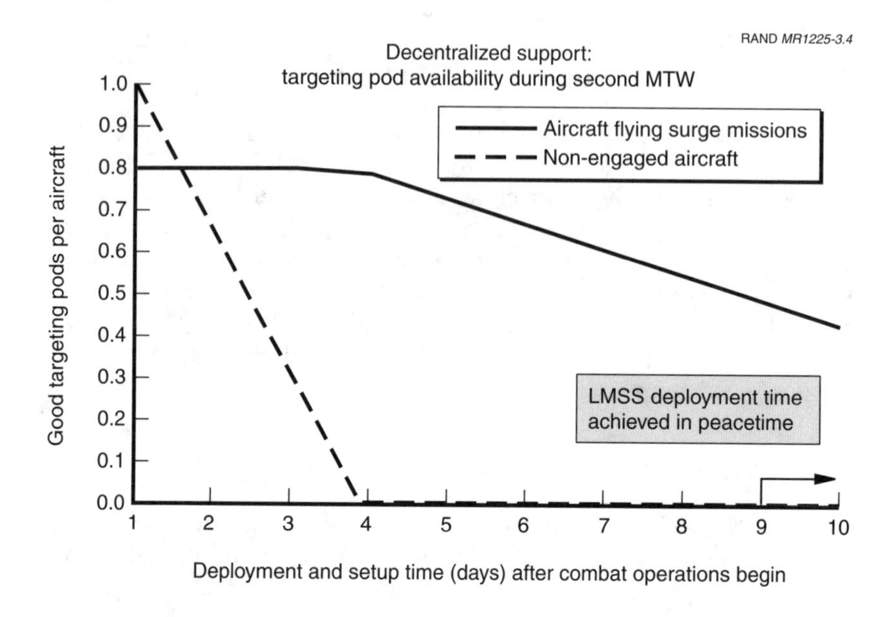

Figure 3.4—Support Equipment Deployment and Setup Time for Decentralized Structure May Affect Targeting Pod Availability

Figure 3.4 shows a pod availability goal of 0.8 combat-capable TRG pod per 1 LANTIRN-capable aircraft during surge operations. This figure exemplifies two decentralized support sensitivities. First, the repair system is sensitive to how quickly support equipment and capability are deployed to the theater. Second, assuming support capability deploys prior to the start of combat operation, if a single tester at an FOL fails, then repair capability at that FOL becomes very sensitive to the supply system's responsiveness in repairing that one tester.

For sustain operations, we set a goal of 0.7 pod per aircraft (with a lower bound of 0.6). With a decentralized structure, spares must be sized to support surge operations because pods cannot be shared across units. Thus, we show availability for aircraft flying surge operations. One reason that we set the ratio of pods to aircraft at these levels is because, given our attrition assumptions, we find that by 2008 the USAF will not be able to meet a goal of 100 percent availability for TRG pods. The maximum two-MTW requirement for targeting pods in 2008 may range from 95 units (for a two-day deployment and setup time) to 256 units (for a 10-day deployment and setup time) above the predicted number of available pods. Since the USAF is not planning on purchasing new LANTIRN pods at this time, we assess how to achieve the maximum use of pods that will be in inventory in 2008 during potential contingencies.

Figure 3.4 shows the effects of deployment times on target pod availability. Specifically, we assumed that during peacetime AEF deployment there would be a ratio of one combat-capable pod for every aircraft deployed. The actual requirement includes safety and pipeline stock for various deployment timelines. In other words, if it takes 10 days to reestablish operations at an FOL, the stock assigned to the deployed units needs to account for surge operations during those 10 days. Once the deployed units' stock requirement was satisfied, we assessed the availability of pods for the non-engaged units.

Because the peacetime with AEF scenario does not significantly stress the support system, we graph the effects of deployment during the second MTW flying a halt operation. The AWOS removal rates modeled were over 50 percent higher than those predicted with current algorithms and varied by aircraft type (see Appendix F). Although the USAF has no policy on TRG pod availability during war, the goal we set for it is above the aircraft availability goal used in calculating RSPs, or 63 percent on day 10 of a contingency. By setting TRG pod availability goals of 80 percent, or well above the aircraft availability level of 63 percent, we ensure that aircraft are not down as a result of pod shortages. To account for scenario variability, we simulated (random selection) removal rates as well as other input parameters to develop 90 percent confidence intervals for resource requirements such as personnel and support equipment. These results will be discussed later in this report.

We assessed the availability of pods for both engaged and non-engaged aircraft. In the two-MTW scenario, there would be approximately 66 aircraft left in CONUS for training purposes. The dashed line to the left in Figure 3.4 shows the expected availability of TRG pods for non-engaged aircraft based on support equipment deployment and setup time, depicted on the bottom axis. Deployment time affects TRG pod availability through the surplus pods to be available until repair capabilities are established. The top of this line shows that we can expect one TRG pod to be available for each non-engaged aircraft if contingency support deployment takes less than one day. For every one aircraft there would be one combat-capable pod to support training operations. The one-pod availability results from our goal of only 0.80 pod per engaged aircraft. If LANTIRN contingency support equipment deployment and setup time takes more than one day, then TRG pod availability begins to decrease. If MTW contingency support deployment and setup time takes more than four days, there will be no pods available to support training missions.

Deployment times beyond four days affect TRG pod availability for engaged aircraft, as shown by the solid line on the right in Figure 3.4. As deployment times extend from five to 10 days, pod availability drops by close to 40 percent for the 378 aircraft engaged in sustain operations for the first MTW and surge in the second MTW. Again, deployment times of less than one day do not affect pod availability for non-engaged aircraft, deployment times from one to four days limit the number of pods available for training aircraft, and deployment times beyond four days affect pod availability for aircraft engaged in contingencies.

Deployment times of the current LMSS are more than 10 days, or long enough to lower TRG pod availability rates below 0.50 good pod per aircraft, well below the level selected to support sustain operations. Investment in an ADK upgrade for the current system, however, may mitigate this deployment sensitivity. ADK deployment is estimated to take just two to three days (assuming that strategic airlift is available on day two of combat operations), while that of the current LMSS is over 10 days. Note that although we used specific ranges in our assessment, the decentralized option is extremely sensitive to deployment and setup time. Because the current support structure without equipment upgrades is expected to take at least 10 days to deploy and set up, it appears that a decentralized structure with no equipment upgrades introduces tremendous risks to warfighter capabilities in a two-MTW scenario.

Although the ADK investment may offer some advantages to the current system, particularly in avoiding the risks of lengthy deployment, it does not eliminate the greater need for spare LRUs that decentralized systems have over centralized systems (see Appendix E). Data from the Warner Robins SPO in 1998 show that LRU depot repair time—the time from when an unserviceable LRU was received at a depot until the serviceable LRU was shipped out—exceeded 30 days. We included transportation times to assess the total loop time from an FOL to the depot and back again. As of June 1999, depot repair times had dropped to 26 days. Because the time LRUs spend awaiting parts accounts for much of the repair lag, we believe that further improvements in repair times may be possible. Still, even if the total loop times were cut to 20 days, we project a substantial investment would be needed for spare LRUs. We estimate (using pre-AWOS data) that $6 million is needed for spare LRUs in a decentralized structure, with or without the ADK investment, whereas only some $250,000 is needed for spare LRUs in a centralized structure. Clearly, spares planning based on the AWOS data will substantially increase the investment requirements (by over 50 percent), although the relative difference between centralization and decentralization should not change significantly. Although new LRU spares are needed for either a centralized or a decentralized structure, the decentralized structure requires a higher level of investment because it has higher safety stock requirements owing to its greater number of locations. The time loops needed for LRU repairs point to an additional sensitivity of time for the decentralized structures.

Centralized Structures

A decentralized system is sensitive to deployment time, and a regional structure is sensitive to transportation time. Just as delays in deployment affect TRG pod availability in a decentralized structure, so delays in transportation between support and operating locations affect availability in centralized structures.

Figure 3.5 shows how transportation times affect TRG pod availability in a centralized structure. We modeled all elements of door-to-door delivery as the total one-way transportation time, including packing, moving, and delivery of pods.

We assess pod availability during the second MTW as a function of one-way transportation time, as shown in the horizontal axis of Figure 3.5. As we did for analyzing time sensitivities in the decentralized structures, we use availability goals of 0.8 pod per aircraft for surge

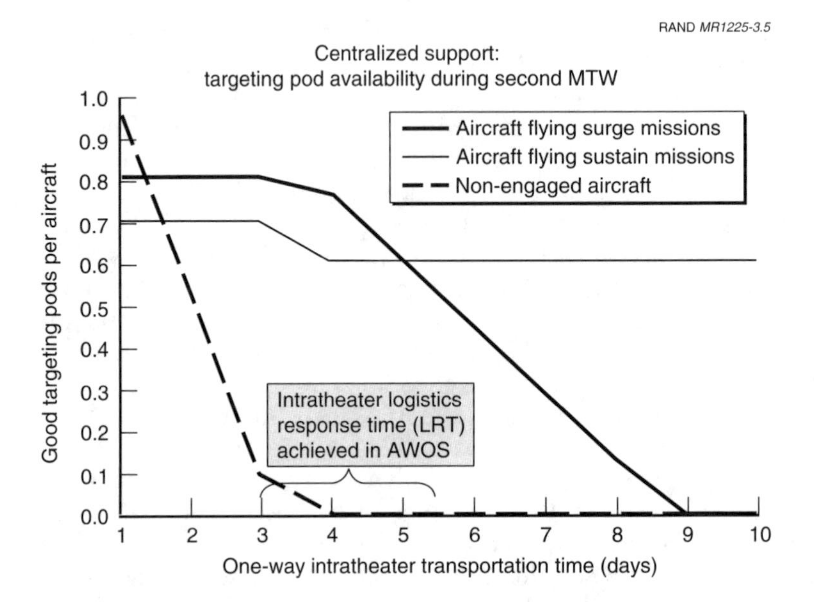

Figure 3.5—Centralized Pod Repair Is Sensitive to Transportation Delays, Affecting TRG Pod Availability

aircraft and 0.7 pod per sustain aircraft (with a lower bound of 0.6). The dashed line to the left shows the expected availability of TRG pods for non-engaged aircraft: non-engaged aircraft can expect one pod per aircraft availability if one-way transportation time for a centralized repair structure is less than two days. If one-way transportation time between operating and centralized repair locations exceeds two days, however, pod availability declines. If one-way transportation time exceeds four days, there would be no pods available to support training missions, and pod availability to the warfighter starts to decline. The thin solid line shows pod availability for aircraft engaged in sustain operations for the first of two MTWs. As transportation times extend from four to 10 days, we allowed pod availability to drop to 0.6 pod per aircraft for the 186 aircraft engaged in sustain operations for the first MTW; nonetheless, availability for the aircraft flying surge operations also declines.

The bracket on the figure shows the average logistics response time (LRT, including backorders) range achieved during the AWOS for inter- and intratheater transport. Additionally, AWOS data indicate that CONUS-to-theater transportation may have greater variability than intratheater movement. This implies that LANTIRN support cannot rely on a CONUS-only support structure because greater pipeline variability will require more pod spares than are available. Thus, relying on support based only in CONUS would result in lower pod availability levels than are needed to wage two overseas MTWs.

Regional support structures composed of a CONUS location and two or more OCONUS locations can be designed to provide one-way transportation times of three to five days between support and operating locations, as demonstrated in the AWOS. However, any regional structure is very sensitive to transportation delays, with transportation times of more than two days affecting non-engaged aircraft used for training and those of more than four days affecting MTW sustain and surge operations.

Footprint

A second performance metric we use in comparing logistics structures and investment options is deployment footprint. We define footprint in terms of the number of personnel and pallets of support equipment that must be deployed to support contingencies. Figure 3.6 shows personnel and equipment pallet deployment requirements for centralized and decentralized repair options

RAND *MR1225-3.6*

	People		Equipment (Pallets)	
Repair Option	Peacetime with 2 AEFs	2 MTW Halt	Peacetime with 2 AEFs	2 MTW Halt
Decentralized (deploy to FOL)	44–48	100–112	30–88	105–374
Centralized (deploy to FSL)	44–48	0–75	0	0

Figure 3.6—Personnel and Equipment Deployment Requirements for Decentralized and Centralized Repair Options

supporting peacetime with AEF or MTW operations using illustrative wartime flying profiles and the AWOS pod removal rates. We show the 90-percent confidence level generated from our simulation model (see Appendix A).

In Chapter Two we discussed the element of time and how a decision to centralize or decentralize repair hinges on the risks the Air Force is more willing to accept. We now shift our attention to the deployment footprint associated with these two options. Deployment footprint affects system performance through its effects on deployment and setup times. The larger a deployment footprint is, the more difficult it may be to attain strategic airlift early in a contingency. We assume that trainees deploy only to support AEF combat operations and so compute the minimum personnel resources required for MTW operations. For boiling-peacetime AEF operations overseas, the current structure may require a deployment of 44 to 48 personnel and 30 to 88 pallets of equipment to support 30 to 40 aircraft per AEF.

Similarly, the centralized structure could require a deployment of up to 48 persons for such operations but no equipment deployment. For halt-phase operations in a two-MTW scenario, the current structure may require a personnel deployment of 100 to 112 persons and an equipment deployment of 105 to 374 pallets.[1] The centralized structure could require a deployment of up to 75 persons (assuming that OCONUS sites are manned at minimum levels with no trainees) for such operations but, again, would require no equipment deployment.

Technology upgrades and consolidation levels account for considerable variation in equipment and personnel deployment requirements. Investment in the ADK upgrade to the LMSS, for example, cuts equipment deployment requirements. Deployment of the current LMSS in a decentralized structure supporting the halt phase of a second MTW may require 374 pallets, whereas deployment of this system with an ADK upgrade could require some 105 pallets. These estimates are based on an 80 percent test set availability rate, which we discuss in the next section. We also show that the total test set requirement is above today's Air Force inventory.

---

[1]The LMSS requires approximately 11 pallet positions for deployment whereas the ADK requires approximately five.

There are three further influences on personnel deployment requirements. First, consolidation cuts personnel requirements by requiring fewer test sets and, hence, fewer personnel. Second, the 60-percent peacetime productivity assumption used to set manning levels greatly inflates the number of personnel required at regional locations. Essentially, this productivity assumption leads the Air Force to buy an "insurance policy" to support potential conflicts. Third, by establishing and enabling USAFE and PACAF regional sites to support rotational AEF forces of 30–40 aircraft, the total number of personnel prepositioned in potential hot zones could increase. Through consolidation, the USAF could manage to keep a substantial number of personnel in theaters of strategic interest but at greater distances from operating locations, thereby reducing personnel turbulence as well as operational and personnel risk during conflicts. Clearly, this must be balanced by the CONUS-to-OCONUS personnel ratios the Air Force needs to maintain. In Figure 3.6, we assume that FSLs are not manned with people to support AEFs; thus, the peacetime personnel deployment numbers are similar to the decentralized structure.

Test Sets and Manning

Not only do the numbers of equipment and personnel to be deployed vary by structure and contingency, so do the total quantities needed. We next give the resources needed by the two main support structure options—decentralized and centralized—to meet the operating requirements across various scenarios. We focus on the test equipment and personnel needed to meet the peacetime-with-AEF and the two-MTW scenarios. Appendix F discusses several key sensitivities uncovered in our analysis as well as a comparative assessment across the range of consolidation options.

Figure 3.7 is an example of how we computed the 90-percent confidence level[2] for test set requirements. Each simulation run of our models generates a curve for the cumulative probability of requiring a given number of test sets. In this case, we show this distribution for

---

[2]A 90-percent confidence level implies that in our Monte Carlo (random) draw simulation, there is a 90 percent probability that the number of resources selected will be able to support the expected demand on the repair shop.

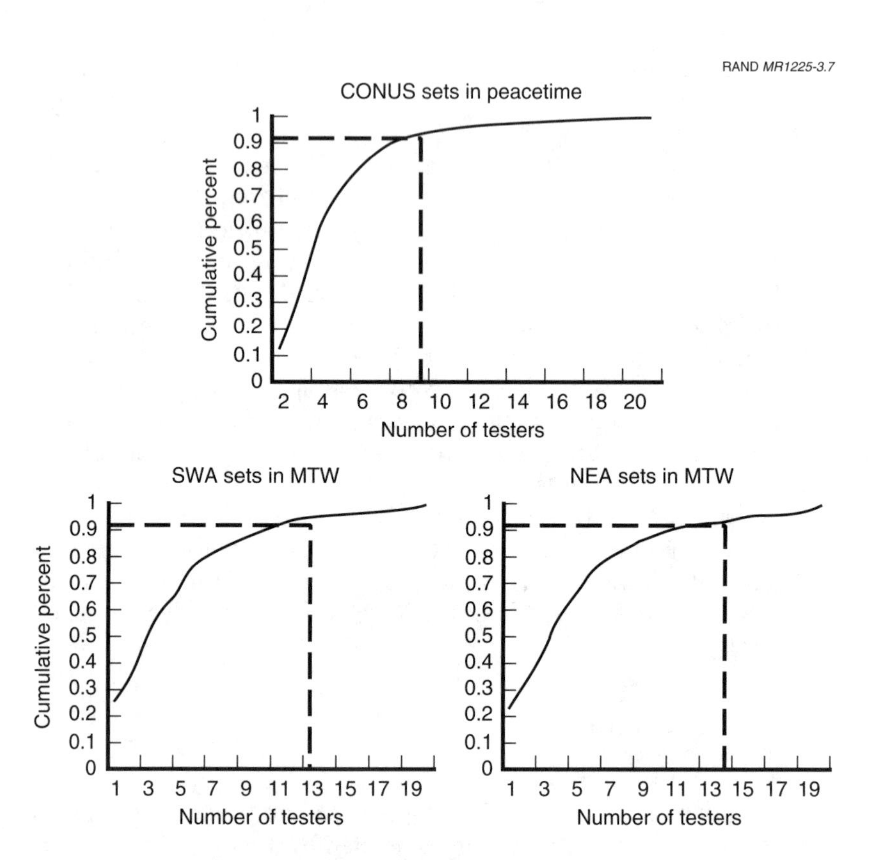

Figure 3.7—Example: Cumulative Distributions of the Number of Test Sets Required at 80 Percent Availability

upgraded equipment during peacetime in CONUS and in two MTWs, one in SWA and one in NEA. The dashed lines represent the 90-percent confidence interval for each region. For example, given an 80 percent availability for the support equipment, SWA may require 14 test sets to have a 90 percent probability of supporting all of the wartime demands. Summing the number of testers in each region, we obtain the global requirement—in this case about 39. This value is again reflected in Figure 3.8, on the right-hand side under centralized, upgraded equipment. A similar approach was used to develop the confidence levels for personnel requirements.

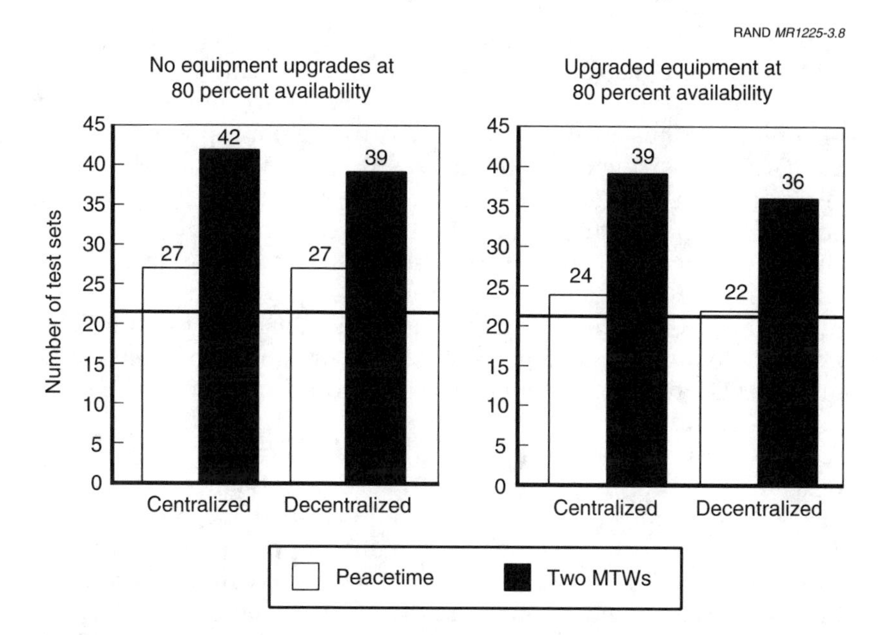

Figure 3.8—Test Set Requirement Across Two Scenarios, Logistics Structures, and Investment Options

Figure 3.8 shows the 90-percent confidence level for the number of test sets needed to support both peace and two-MTW-wartime operations given today's support equipment or the upgraded version, and AWOS removal rates. We show expected requirements at an 80-percent availability level and discuss sensitivities to this assumption in Appendix F.

Three observations can be drawn from these figures. First, investing in support equipment upgrades does not significantly reduce the number of test sets needed, given an equal availability level. Predicting the performance of either today's equipment or that with upgrades is very difficult, so we assessed the sensitivity of this metric and found only minor changes to the overall requirement quantities. Thus, the largest potential benefit to upgrading the support equipment may be in improving the deployment footprint, as discussed earlier. Clearly, if the Air Force chooses to centralize repair operations, deployment footprint becomes a negligible point.

This leads us to the second point made by the figures. Shifting from a decentralized to a centralized support structure may not necessarily reduce the total number of test sets needed to support wartime operations because most sets in CONUS would not support the effort and OCONUS sizing needs to include AEF deployments. However, centralization reduces repair operation sensitivities to support equipment spare shortfalls, as discussed earlier. Additionally, we show in Appendix F that other consolidation options may offer economies of scale and hence resource requirement reductions. Again, this implies that the most significant advantage to centralization is reduction of the deployment footprint, and as demonstrated in the AWOS, this can benefit combat support. Finally, and most important, Figures 3.7 and 3.8 indicate that the Air Force is underresourced to support two coincident MTWs. Across all options, the peacetime requirements are for 22–27 sets. With a current inventory of 21 testers supporting combat-coded units (solid line in Figure 3.8), the Air Force may have enough equipment for peacetime operations. However, the two-MTW scenarios modeled indicate that the total inventory may need to increase by over 50 percent to ensure warfighter support. With new testers costing close to $20 million each, this finding has broader cost and planning implications. If the Air Force (and the Defense Department) want to continue planning for two coincident MTWs, there may need to be a significant investment to ensure support of LANTIRN resources. Alternative remedies could be to rely more heavily on other PGM technologies or to revise the WMP and DPG scenarios.

The total personnel requirements across this decision space are not as daunting. Figure 3.9 shows 90-percent confidence levels for the number of people required across the options discussed above, with an expected test set availability of 80 percent. We computed wartime personnel requirements (excluding trainees) to reflect minimum manning numbers. The peacetime bars reflect both direct labor personnel and trainees. Again, there is little difference between investment options and logistics structures. Furthermore, our models indicate that there may be sufficient people to support multiple contingencies (current assigned total is 307—see the solid line). Again, this implies that the primary tradeoff between centralization and decentralization hinges on deployment and transportation capabilities.

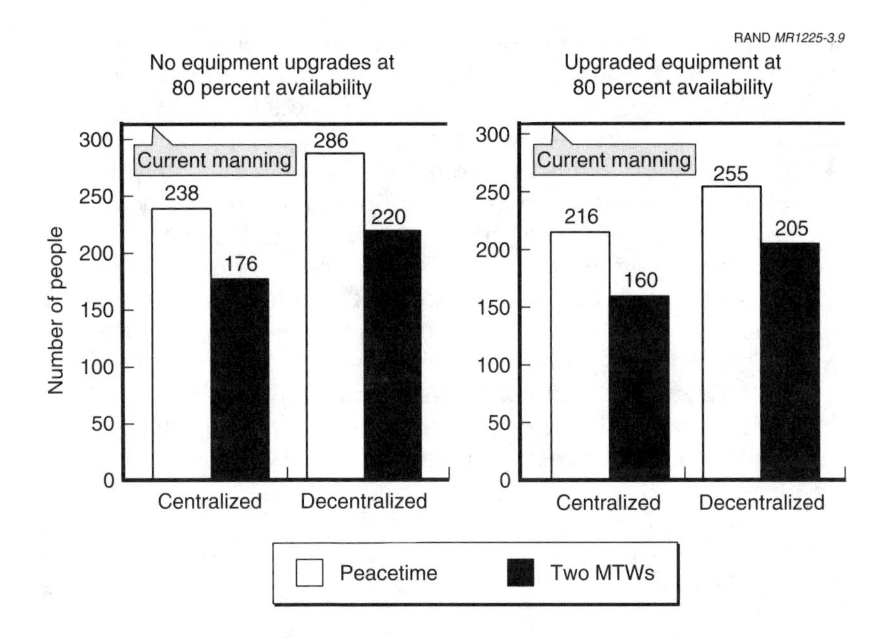

Figure 3.9—Personnel Requirement Across Two Scenarios, Logistics Structures, and Investment Options

Clearly, incorporating skill-level degradation and trainees for combat operations could increase this number somewhat. Because modeling these effects is difficult, we assessed the relative sensitivity of our outcomes to pod repair time (one measure of operator skill level). This metric offers a measure of how well people work, in other words, the number of pods a given crew can process. We discuss this sensitivity in Appendix F.

Recurring and Investment Costs

The performance metrics we have reviewed indicate that new investments are required to upgrade the current equipment to meet the support needs of a two-MTW halt-phase scenario. One of the metrics we need to consider for LANTIRN support structures is the new investment required for equipment upgrades, infrastructure, spare LRUs, and personnel relocation. Investment costs are not the only financial variable to consider when evaluating systems. Each structure carries recurring costs for peacetime labor and transportation. We consider both recurring operating costs and nonrecurring investment costs using expected cash-flow models. We discuss the calculations for our present-value analysis in Appendix H. As pointed out earlier, our assessment of AWOS data indicates that the Air Force may need to increase its equipment inventory by over 50 percent. Because we were unable to obtain accurate acquisition costs for new support equipment, we did not model the expected costs of purchasing additional testers. The relative recurring cost differences across all the options are negligible given a fixed set of inputs. Appendices F and H offer a comparative analysis of the options analyzed, including the AWOS removal rates, and illustrate this point. Again, the Air Force's decision should focus on expected capabilities and risks rather than the costs of the various options.

Chapter Four

EVALUATING THE OPTIONS

How should the USAF choose among the options available to it for LANTIRN support? What are the major issues in choosing between a decentralized or centralized structure? What is the proper level of investment to make for the future support structure? The answers to these questions depend largely on what characteristics USAF decisionmakers most value.

OVERALL COMPARISONS OF CENTRALIZED AND DECENTRALIZED STRUCTURES

Figure 4.1 compares the pros and cons of the two overarching options we analyzed. Although the centralized option may require fewer test sets and personnel, its annual transportation costs may be higher. As a result, the combined recurring costs for decentralized and centralized structures may be approximately the same.

The regional support structure does offer a reduced deployment footprint advantage. Consolidation cuts or even eliminates personnel and equipment deployments needed to support contingencies. Furthermore, because FSLs are removed from theater operations, consolidation reduces risks to both support equipment and personnel. Although regional operations may become more vulnerable to attack (both conventional and cyber), proper preparations and communications design can alleviate these threats.

The collocation of test equipment with consolidation would reduce single-string risk and eliminate the need to transport repair equipment in supporting contingencies. Consolidation would also elimi-

RAND *MR1225-4.1*

	Decentralized	Centralized
Number of test sets	Higher	Lower
Number of highly trained personnel	Higher	Lower
Operating costs	Neutral	Neutral
Deployment footprint	Higher	Lower
Single-string risk	Higher	Lower
Deployment transportation risk	Higher	Lower
Resupply transportation risk	Lower	Higher
Pod transportation risk	Lower	(a)
Logistics command and control system	Lower	Higher
Investment	Higher	Lower

[a]Air National Guard has been able to eliminate this problem.

Figure 4.1—Logistics Structures and Risks

nate deployment transportation risks, particularly the risk resulting from lengthy deployment times for equipment transport, setup, and proper functioning in the theater. Furthermore, consolidation mitigates the risk of a single string going down and temporarily halting repair operations at an FOL during combat operations.

However, daily pod transportation risks increase with the consolidated option. Because consolidation requires pods to be moved off base for repair, system performance becomes sensitive to transportation delays. Consolidation may require pods to pass through more transportation channels, involving more personnel in the process. Rough handling in the new channels may become an issue in the proposed regional structure. Although our discussions with the Tulsa Air National Guard indicated that pod handling during transport is not a major issue, standardized training for all personnel handling pods may be advised.

Most important, the consolidated intermediate repair structure will require new organizational processes. Unit commanders will have to relinquish some of their control over LANTIRN pods and communicate closely with the support centers and other bases serviced by the same regional facility. Conversely, a decentralized structure may be more heavily dependent on a robust logistics command and control system to ensure timely resupply of spare parts to repair both pods and support equipment. Thus, in both cases, performance metrics and incentive systems may need to change to support a system focused on customer (warfighter) satisfaction, on-time delivery, and quality workmanship.

Finally, although we suggest that the relative operating costs between each of the options is negligible, the Air Force may be under-resourced to support multiple large-scale contingencies. This finding has a much greater cost implication because, regardless of the support structure chosen, a significant investment may be necessary to ensure certain capabilities.

DESIGNING AND TESTING A NEW LANTIRN SUPPORT STRUCTURE

Our analysis has shown that both centralization and decentralization offer opportunities for improving today's intermediate-level maintenance of the LANTIRN system. Furthermore, the Air Force's experience in the AWOS has highlighted support capabilities and limitations in a wartime environment. Centralized repair was formalized for LANTIRN pods employed in the AWOS. We offer for consideration here a centralized three-regional LANTIRN support structure using one CSL and two FSLs. Figure 4.2 shows a notional beddown of support equipment and personnel for a three-regional structure with an ADK or MLU upgrade.

At the 90-percent confidence level, 10 sets would be placed in CONUS, 15 in PACAF, and 14 in USAFE. The number of test sets in PACAF and USAFE are selected to support a halt-phase MTW in each region. The 10 sets in CONUS would support peacetime operations as well as training missions during the MTWs. Note that individual bases (or FOLs) would have virtually no I-level testing capability.

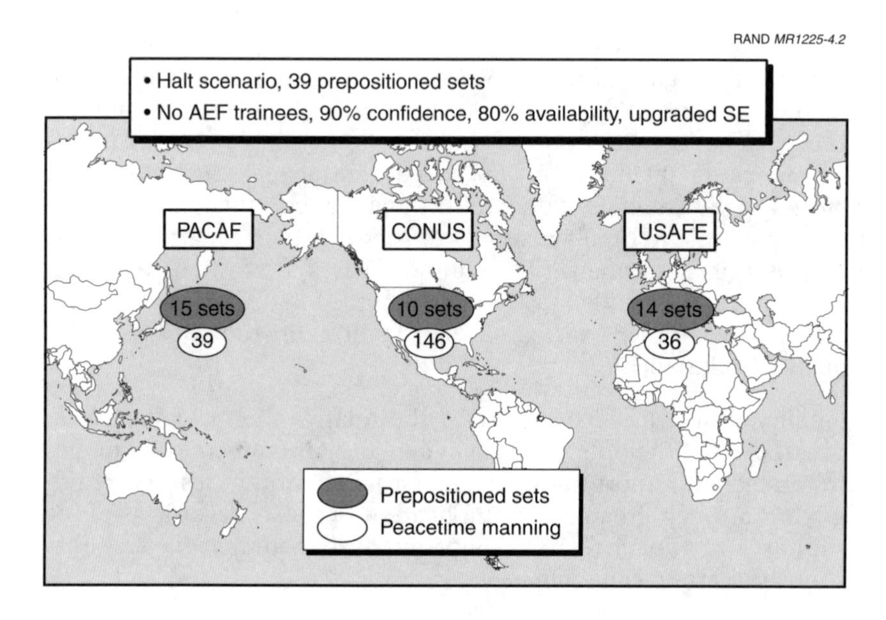

Figure 4.2—Proposed Positioning of Testers and Personnel

They would have no LMSS (type) support equipment and no I-level personnel. These bases would ship pods to an FSL.

The manning numbers indicate the minimum number of personnel in each FSL during peacetime. Note that the OCONUS manning does not include trainees. The CONUS site would accommodate additional trainees who could augment wartime and AEF deployments.

Note that this structure may require up to 39 test sets—18 more than are currently in inventory. Recall that the test set and personnel computations are based on AWOS removal rates and WMP flying profiles. We assume that both NAV and TRG pods are supported. Given that our AWOS data are based on only two units flying LANTIRN during the war, we applied our removal-rate estimates through Monte Carlo simulations. These results may seem somewhat overstated, but they nonetheless reflect a potentially significant support capability shortfall.

CONCLUSION: TESTING AND EVALUATING A NEW SUPPORT STRUCTURE

Although a system relying on a CSL in network with FSLs introduces new transportation time risks, we conclude that such a system offers distinct advantages over the current system, particularly when combined with an Advanced Deployment Kit or Mid-Life Upgrade that can potentially improve equipment reliability. The most viable structure we identified would use two FSLs and one CSL, all with the new technology upgrades. This option consistently ranks high when all options are considered by cost and performance measures, including pod availability, reduced deployment footprint, or present value of recurring costs. Although the underlying premise of the FSL-CSL network is that no equipment moves in support of deployed units, the ADK offers additional flexibility to accommodate scenarios that the fixed support structure could not.

During the Air War Over Serbia, the U.S. Air Force employed some of the centralized repair concepts proposed by our research. Fighter aircraft deployed to an FOL in Italy received LANTIRN support from their home base at Lakenheath. No LANTIRN support equipment deployed to the FOL and pods were transported via multiple modes, enabling responsive support. Although this limited experience did not fully stress the LANTIRN centralized support system, it provided insights to the potential feasibility of such a system. Based on our work prior to the AWOS and the lessons learned from the war, we recommend that the U.S. Air Force further explore the benefits and risks associated with LANTIRN repair consolidation. As a first step, we recommend that the Air Force invest in the ADK to ensure continued repair capabilities and improve deployment flexibility. Both our research and the AWOS have shown that a transportation system able to respond to a wide variety of scenarios ranging from peacetime to two coincident major theater wars is necessary for successful centralized repair operations. Thus, our second recommendation is that the Air Force reevaluate the capabilities of its intratheater transportation systems, starting with the command and control processes used to manage materiel movement. Only after gaining a solid understanding of the transportation system capabilities can the Air Force pursue implementation plans for centralized repair structures.

Appendix A

MODEL FLOW CHART AND COMPUTATION EXAMPLE

Our study methodology integrated three distinct models, applied through a systematic evaluation process. Figure A.1 depicts the overlying analysis process and model integration points. Beginning at the upper left-hand corner of the chart, we calculated the peacetime LANTIRN average removal rates to the back shops using form 095 pod logbooks from Moody, Mountain Home, and Elmendorf Air Force Bases. Then, applying extrapolation algorithms, we estimated the wartime removal rates. The underlying premise of these algorithms is that sortie frequency, not duration, drives removal rates. For example, as sortie rates change from one to two sorties per day, the pods removed per sortie increase by about 9 percent, almost completely independent of the sortie length. We assessed option outcomes relative to these predicted removal rates as a basis for our comparative strategic decision space. However, to evaluate wartime removals more accurately, we based our computations (presented in the body of this report) on data collected from the AWOS. Appendix F outlines our findings and recomputed wartime removal rates from this latest operation. We modeled the possible range of these rates using a Monte Carlo simulation in which we varied the mean[1] removal rate randomly in concert with several other probabilistic inputs discussed later in Appendix A.

---

[1]We address future scenario uncertainties by randomly varying the mean values of removal rates, test set productivity, and repair time to account for potentially large swings in our predicted values. The distribution ranges used were based on data collected from operating bases as well as the AWOS. We account for local variations around the mean through a 95th percentile calculation for stockage requirements (shown in the example).

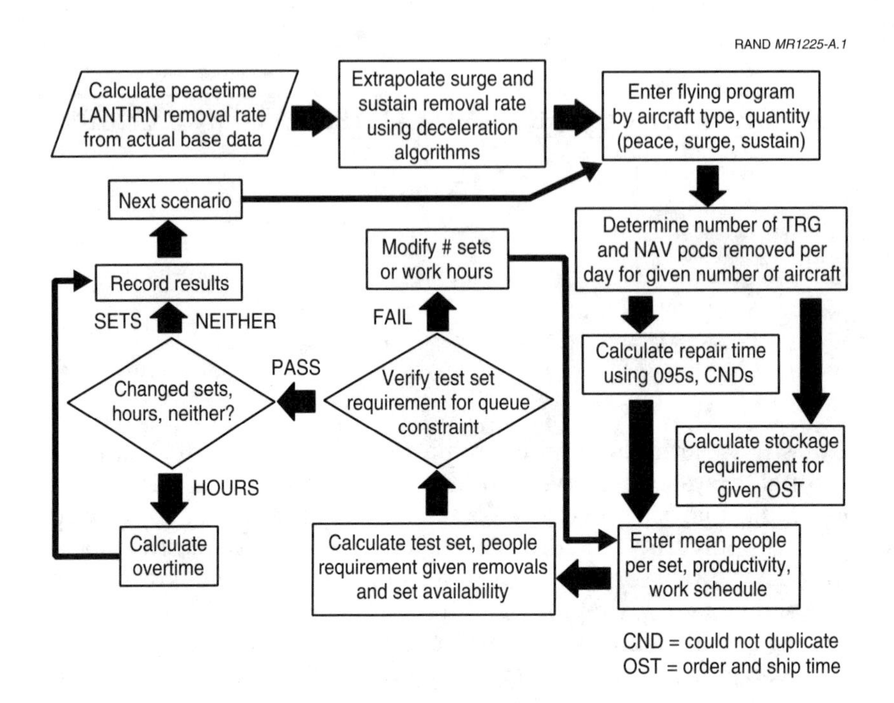

Figure A.1—Analysis Process and Model Integration Points

The next set of inputs includes the number and type of aircraft that we wish to model, as well as the type of flying program the aircraft will fly. We assumed aircraft were flying at either peace, surge, or sustain rates. The model automatically selected the appropriate removal rates for the flying program. The model also calculated the number of expected pods removed per day for a given location. The number of pods removed supports several computations in the next module. First, each pod corresponds to an associated distribution for repair hours. These hours are used for personnel, stockage, and support equipment requirement computations. Pod removal data were fed into two separate spreadsheets. One sheet calculated the stockage requirement based upon transportation pipelines, and safety stock requirements. The analyst can also set specific pod availability targets to mitigate systemwide pod requirement shortfalls.

The support resource determination model requires several other inputs before calculations are run. First, pod repair times are randomly selected from a discrete distribution, and CND ("could not duplicate") times are entered for both targeting and navigation pods. Second, manning per shop, productivity rates, and work schedules are entered. For example, there may be five persons per shop for each of two eight-hour shifts, working at 60 percent productivity.[2] Finally, equipment availability as a function of collocated sets was calculated separately. The analysis included a regression model[3] as well as daily hours of operation and work schedules.

Given these supply-and-demand relationships, the model selects the number of repair strings required and the associated personnel. The model also calculates workspace requirements and the peacetime loading for the pod transportation analysis. Finally, the model predicts the annual direct-labor operating costs for the location specified. Before recording these results and moving on to the next location or scenario, the analyst must verify the in-shop queue constraints in a third model.

The queuing model used in this analysis assumed that pods arrive at the shop according to a Poisson distribution. They can be repaired by multiple test sets depending on shop resource constraints. In this model, the repair time is exponentially distributed. The model output is the expected number of hours a pod would stay in the repair shop before being serviced. The analyst must use this output to determine if queue thresholds have been surpassed. For example, if the

---

[2]Appendix H discusses the productivity assumptions used.

[3]Because of lack of data on the effects of collocating LANTIRN test sets, we employed a regression curve similar to that developed in an associated RAND report addressing support equipment collocation options (see Eric Peltz et al., *Supporting Expeditionary Aerospace Forces: An Analysis of F-15 Avionics Options,* 2000). The equation used in the LANTIRN models is:

$$C = 17.82 * A - 0.00078 * B + 1.90 * \ln(B)$$

where A = expected single-string availability

B = number of collocated strings in a shop

C = average expected available hours per string.

Our Monte Carlo simulation assumed a fixed value for A per given scenario while randomly varying the expected load on the repair shop, thus driving tester requirements.

queue time is over 20 hours, one may change the hours per shift or the number of strings used in the model. Once all constraints have been met, the results are manually recorded in a tracking spreadsheet and the modeler can proceed to the next set of scenario and location assumptions. When running a scenario simulation with the randomly selected inputs described above, the output is in the form of a probability distribution as shown in Figure 3.7.

COMPUTATION EXAMPLE

An analyst would compute the resources required to support LANTIRN targeting pods on a squadron of 18 F-16 aircraft deployed on a combat mission as follows. The mission profiles we assumed have a maximum sortie rate early in the program (say day 15 of a 50-day scenario). We would then select day 15 upon which to base our resource requirement computations. Now, given a mean wartime removal rate of 0.04 TRG pod per aircraft per sortie per day, the expected number of pods removed on a given day is 18 * 0.04 = 0.72 pod. When running the Monte Carlo simulation, the model would generate a distribution of expected pods per day. Next, to compute the number of expected hours required to repair the pods, the model multiplies the number of pods by the expected repair time. Using the mean powered-on repair time for TRG pods, we first compute the total repair time by increasing the on-time by 30 percent to account for powered-off processing. The interim value becomes 13.8 (hr) * 1.3 = 17.94 hr. Next, we must account for processing time when no repair took place or for CND time. This process has a mean time of about 5.5 hr. We also found that about 7 percent of all TRG pods do not need repair. The expected repair time then becomes 17.96 * (100% – 7%) + 5.5 * (7%) = 17.09 hr. But before we continue, we wish to account for potential repair-time improvements resulting from upgraded support equipment. Suppose the mean repair-time improvement is 13 percent, then the expected repair time becomes 17.09 * (100% – 13%) = 14.9 hr. Again, our simulation varied repair time and repair-time improvement to yield a probability distribution. Given this mean repair time we now compute the expected load on the shop as 0.72 pod * 14.9 hr = 10.73 hr per day.

Clearly, this example puts minimal loads on the shop, yet we still must compute the minimum resources required. First, to compute personnel requirements, we assume that a minimum of two people are required per shift for every fractional or whole pod expected to arrive in the shop. Because we are modeling a wartime scenario, we set the number of shifts at two, at 12 hours each. So in our case a minimum of 2 * (round up of 0.73 pod) = 2 people are required to process the pods. Next, we must modify this number by the expected productivity rate for wartime operation (we used 90 percent). The number of people becomes 2/0.9 = 2.22 direct labor personnel. Now we compute the overhead labor requirements. We start with one supervisor per shift, which in our case equals two. Then we add one shop chief per shop for a total of three overhead personnel.

Before we compute the trainee and supply personnel requirement we must first calculate the number of test sets required for this example. As described earlier, we expect about 10.73 hours of repair time per day. Given a 24-hour-per-day operation, we may expect to have equipment available 24 hours, but this availability must include the expected performance of the equipment. Because daily performance data were not available, we used an average mission-capable (MC) rate metric. Forecasting MC rates is difficult, so we chose a range of possible performance levels (see Appendix B). If we have an MC rate of 80 percent, then the expected available hours per a single tester would be 24 * 80% = 19.2. Next, we would equate this value to the expected workload of 10.73 hours resulting in a requirement for one string. These values are then fed into a queuing model (described earlier) with which the analyst can verify the expected wait time for each pod; in this case it is 12.04 hours. Supporting one tester with upgrades requires about one-half of a full-time person, so now the total shop manning becomes: 2.22 direct personnel + 0.5 supply person + 1 supervisor = 3.72 people per shift, or 7.44 people per shop + one shop chief = about 9 people. One can also add trainees to the total manning as a function of the number of test sets available for training.

Finally, we want to compute the spare pods required to support the 18-aircraft squadron. Starting with a removal rate of 0.73 pod per day, we need to decide on an availability goal. In our analysis, we used an 80 percent availability for engaged aircraft. In other words, we would like to maintain 0.8 good pod for every aircraft. So in our example we would need a minimum of 0.8 * 18 = 14.4 pods. The expected number of pods that would typically fill the repair pipeline will increase this figure. Suppose it takes two days to repair a pod and about one day to transport it from the flight line to the repair shop and another day back. The total repair pipeline becomes 2 + 1 + 1 = 4 days. Next, we compute the number of pods that may be removed during this time: 4 days * 0.73 pod per day = 2.92 pods. To this we add a 95-percent confidence level to account for process variability by computing the square root of 3 * pipeline = 2.96 pods. The total pods needed to fill the pipeline is 2.96 + 2.92 = 5.88 pods. We now add this figure to the minimum good pods required, for a total of 5.88 + 14.4 = 20 pods for the 18 aircraft. So the total resources required to support this unit are one test set, 20 pods, and nine people.

Appendix B

TEST SET AVAILABILITY

To better understand test set degradation, we assessed tester availability. Figure B.1 shows the average monthly LANTIRN Intermediate Automatic Test Equipment (LIATE) mission-capable (MC) rate for test strings at five bases. The bases were chosen because they represent locations with combat-coded aircraft that could potentially deploy for contingencies and have used the same test sets over the three-year period charted. We show LIATE MC rates because this subsystem's performance drives overall LMSS MC rates. The dashed line represents bases with singe test sets and shows no marked per-

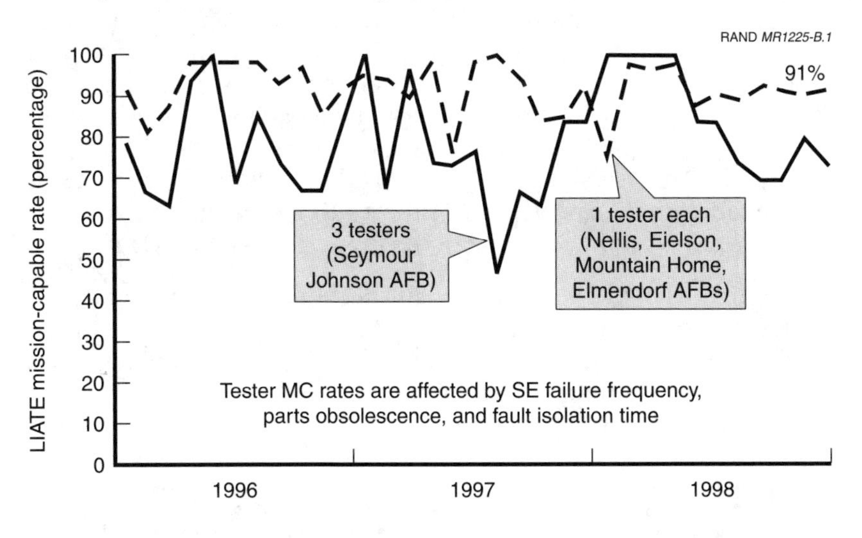

Figure B.1—LIATE Tester MC Rates Over Three Years

formance degradation over three years. The solid line represents Seymour Johnson AFB data where three sets operated in concert and the line also shows no degradation. Note that this base shows much greater MC rate variation and typically performs below the single bases assessed. Although we expect that collocated test set average performance would be greater, our detailed examination of the data indicates otherwise. We believe that one of the main issues driving this anomaly is that all data represent peacetime operations. Thus, whereas single-set bases have to maintain their one tester at peak performance at all times, bases with additional testers may not have as high an urgency, in that not all sets are required to support peacetime operations. The effect is that collocated test sets may stay unrepaired for longer periods of time, driving MC rates down. Because the single-string bases represent achievable performance levels, we used the latest MC rate (90 percent) for our baseline assessment and extrapolated collocated test set performance using relationships developed in an associated RAND study.[1]

Tester availability is driven by three major elements: test set failure frequency, fault isolation time, and availability of spare parts to repair the failure. Test set failure frequency and fault isolation time data were not available, so we looked at overall tester MC rates and forecasted technician resources data to assess possible future equipment performance levels. Operator skill levels most affect fault isolation time; the more skilled the operator, the less time needed to isolate equipment problems. Because the most highly skilled sensor personnel are leaving the Air Force at an increasing rate (shown in Figure B.2), the overall skill level of LANTIRN technicians is decreasing, leading us to expect longer process times. However, we could not draw a direct correlation between these trends and MC rates.

Given that the data do not show a distinct support equipment degradation trend, we developed a matrix approach to possible future equipment performance with no investment and the upgrades described in the body of this report. Figure B.3 shows the two

---

[1]See Eric Peltz, Hyman L. Shulman, Robert S. Tripp, Timothy Ramey, Randy King, and CMSgt John G. Drew, *Supporting Expeditionary Aerospace Forces: An Analysis of F-15 Avionics Options*, RAND, MR-1174-AF, 2000. See also footnote 3 of Appendix A.

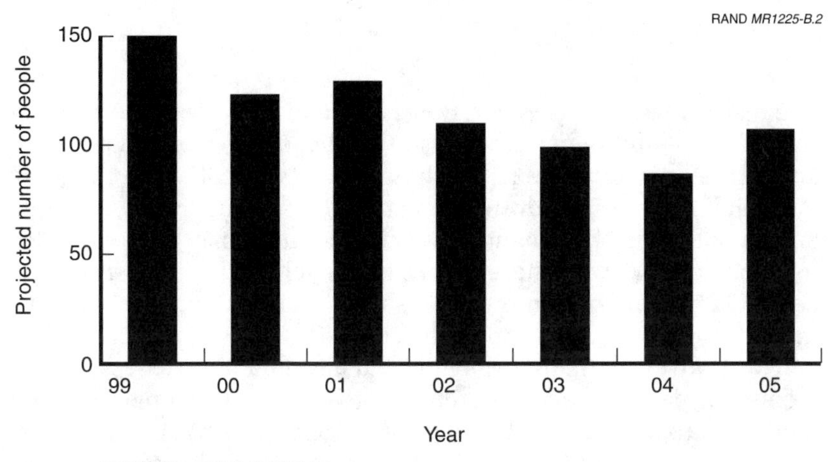

SOURCE: AFPOA/DPYE.

Figure B.2—Projected Numbers of Sensor-Coded Personnel (> 10 years of service)

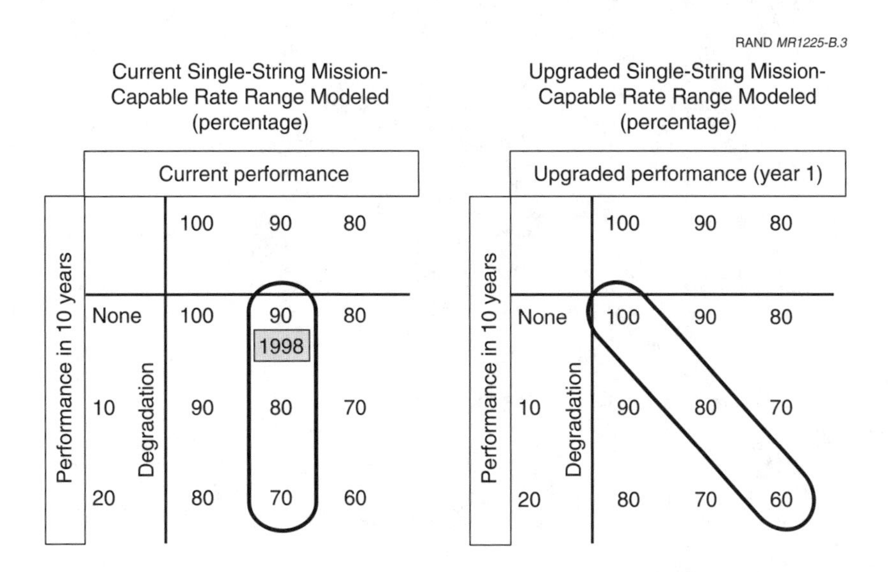

Figure B.3—Possible Performance of Support Equipment

matrices we used to assess the effects of future support equipment performance.

The matrix on the left shows a range of 100 to 80 percent MC rate for the current equipment. We had shown that single test sets perform at about 90 percent. Next, we show on the vertical axis a range of potential degradations over the next 10 years, from 0 to 20 percent. So, the potential performance levels in the future may range from 90 to 70 percent. We use these ranges in our sensitivity analysis in Appendix F. The matrix on the right shows a similar approach for the upgraded equipment. Because the cost differences between an ADK and MLU investment are negligible and it is difficult to assess the incremental performance difference between the two upgrades, we model the new equipment using one range of possible performance levels. At the upper limit, the new equipment may perform at 100 percent in the first year—it is difficult to measure this starting point because few data are available. Again, on the vertical axis we show a range of possible degradations over time. Now, the possible performance range becomes 100 to 60 percent. We discuss the implications of this range in Appendix F.

Appendix C

STEP FUNCTION OF SUPPLY CAPACITY

Figure C.1 depicts how consolidating support equipment affects supply capacity. Suppose, for example, that repair shops operate 24 hours per day with unconstrained personnel availability. In this case, the primary support limitation becomes the number of test sets. If these strings were to operate independently, then one squadron would need a whole string even if it needed only 75 percent capacity of the string, because strings cannot be separated into units

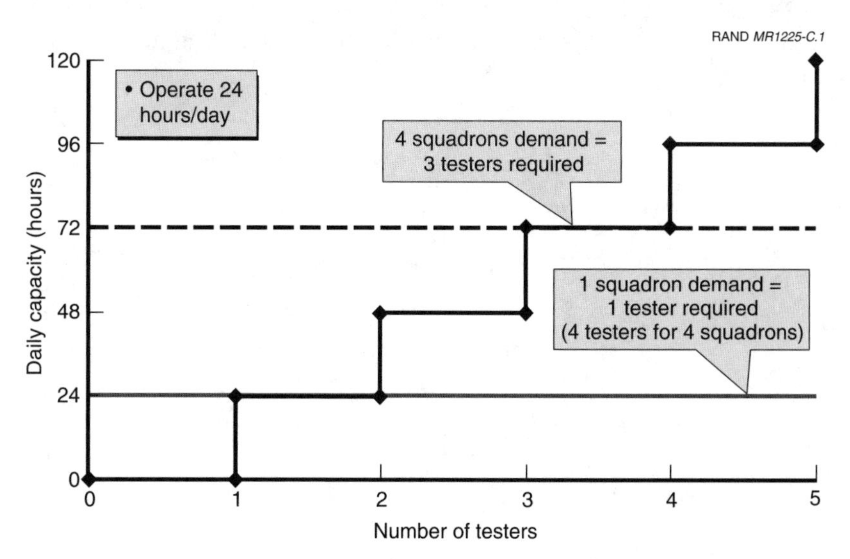

Figure C.1—Supply Capacity Increases with Consolidation

smaller than one. Four independently supported squadrons thus would require four sets of repair equipment.

If, however, the strings supporting these four squadrons were to be collocated, then only three strings, not four, would be required, because the 75 percent string capacity needed by each of the four squadrons could be met by three collocated strings running continuously.

Support capacity is further enhanced by increased uptime per group of collocated sets. As the number of collocated strings increases, their average availability increases as well. Another benefit to collocation is lower repair capability uncertainty once a test set becomes non–mission capable. If a single string at an FOL breaks down during the peak flying period, the availability of support equipment spare parts (for repair) becomes critical. Furthermore, supply system capabilities may also be stressed during surge operations. In a high demand with low resource availability scenario, assured equipment performance becomes extremely critical. Collocation does not eliminate the possibility of equipment failure, but it does provide more certainty that minimum repair capability can be maintained at all times—the chances of all collocated test sets failing at the same time are very low.

Appendix D

VIRTUAL REGIONAL CONCEPT

One variation on the regional concept is what we call a *virtual regional* structure. Figure D.1 depicts a support structure consisting of a CSL with two FSLs with an ADK (or MLU) upgrade investment. Earlier we showed that this structure may require 39 test sets, with 10 of these positioned in one CONUS location. The virtual regional structure would position all 39 test sets in two CSLs with excess ca-

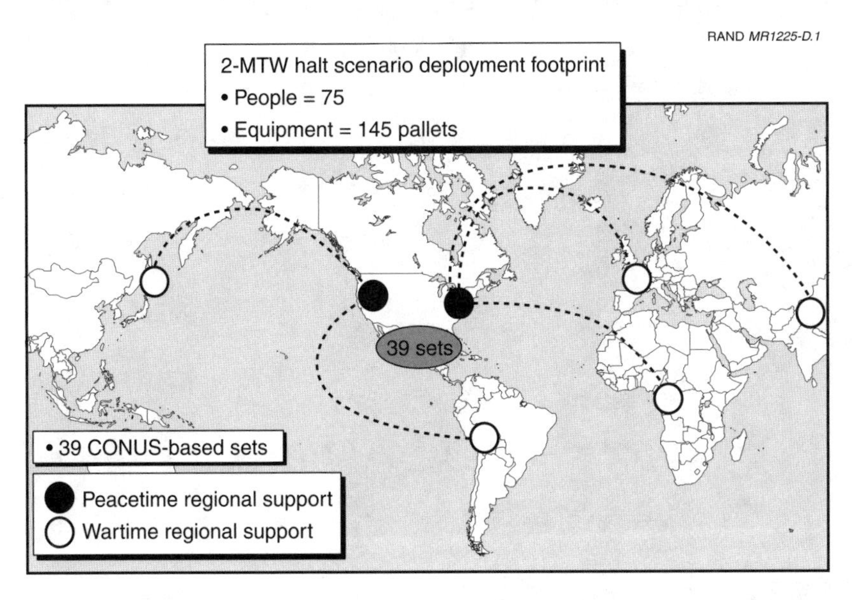

Figure D.1—Structure and Deployment for the Virtual Regional Support Option

pacity. Some of the CONUS sets would deploy with the units and form a regional repair facility near or at the newly established FOL. For example, if a three-squadron AEF to PACAF requires one set per squadron yet only two sets to support all three, then the Air Force could deploy just two sets and form a mini-regional repair center in PACAF. This virtual regional option offers greater flexibility to respond to various contingencies (i.e., flareups far from a fixed regional site), but it has deployment risks for both equipment and personnel. As shown earlier, deployment delays can severely hamper the warfighter's ability to achieve mission requirements. Again, the decisionmaker's preferences can weigh the risks and benefits of this alternative.

Appendix E

LOGISTICS STRUCTURES AND LRU SPARES REQUIREMENTS

Although decentralized support with equipment upgrades offers some advantages, spare LRU requirements must be included when comparing this option with centralized support. For this assessment, we assumed that additional pods are not purchased. Starting with the SPO pod availability goal of 85 percent in peacetime and our 80 percent goal in wartime, we calculated the LRU investment requirement across the three primary operational scenarios. We used pre-AWOS removal rate data and focused on the wartime requirements during a two-MTW halt-phase operation. As discussed earlier, cost comparisons across options are not significant, so the main point of our discussion is the relative difference between centralized and decentralized spares requirements. Because additional LRUs may be purchased, we split the investment requirement into two elements. First, using a 1998 snapshot of serviceable and unserviceable LRU inventories, we assessed the total repair costs needed to bring the required number of LRUs to operating condition. Next, we calculated the number that would need to be purchased to support each scenario if there were insufficient LRUs available systemwide. We used a total loop time range of 15–30 days for the time associated with depot repair processes plus transportation and order processing.

The curves in Figure E.1 represent LRU investment as a function of total loop time for the MTW scenarios. The solid curve corresponds to a decentralized system during the MTWs and the dashed curve represents centralized support requirements. Note that decentralized support requires more than two times the investment in LRUs to support a two-MTW scenario. Data from the Warner Robins

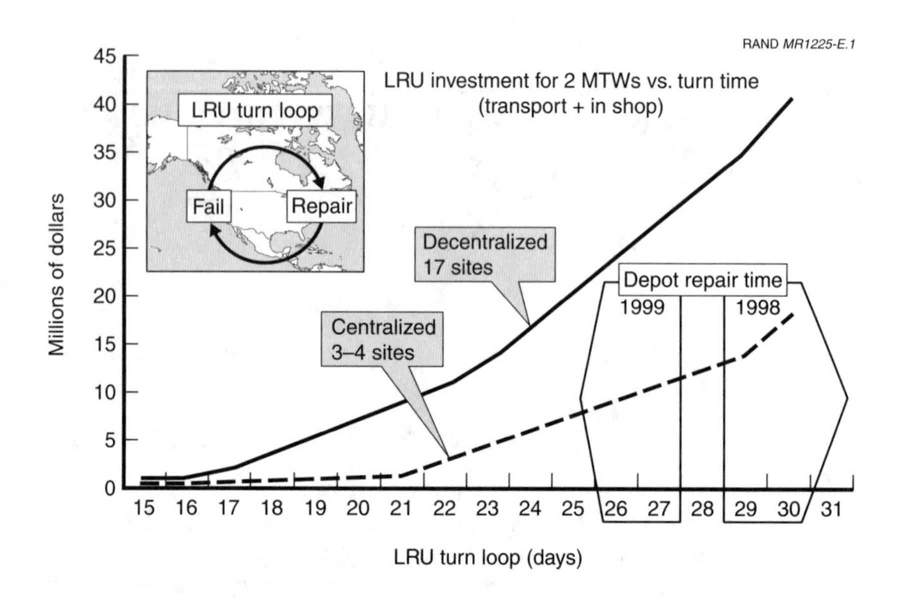

Figure E.1—Logistics Structures, Repair Turn Loops, and Spare LRU Requirements

SPO in 1998 indicate that depot repair time alone averaged about 34 days, so we can expect that the total loop time was over 40 days. This point would fall well beyond the rightmost extreme of the figure. As of June 1999, depot repair times had dropped to about 26 days, or to a point within the 1999 Depot Repair Time shown on the chart.

Furthermore, depot repair process data collected during the AWOS indicate that process times can be significantly lower given adequate resources and prioritization rules.

In assessing a decentralized structure, regardless of SE investment, we estimated that between $5 million and $25 million may be needed in LRU spares investment depending on supply system performance. Centralized support may require between $250,000 and $10 million in LRU investments. These values may be significantly understated because we did not apply AWOS removal rates for this computation. Although the centralization option investment appears lower, recall the sensitivity of this logistics structure to transportation times. If pipeline times are not carefully controlled, insufficient LRU and pod spares inventories will significantly affect pod availability.

Appendix F

SENSITIVITY ANALYSIS

With any model-based analysis, outputs are only as good as the inputs. We collected extensive data from multiple Air Force sources and checked the nominal values used to minimize errors. Prior to designing the analysis model, the LANTIRN working group team verified many of the critical values used. This team, led by RAND, included representatives from Air Combat Command, Warner Robins SPO, Air Force Logistics Management Agency, and HQ USAF.

We realized, however, that no amount of data could fully compensate for the significant uncertainties in future deployment and employment scenarios. We therefore built models to assess input variable uncertainties and augmented our earlier sources with data collected during the AWOS.

Our treatment of pod removal rates from aircraft demonstrates how we accounted for uncertainties. Aircraft pod removal rates are the primary influence in determining aggregate repair asset requirements. We were able to collect reasonably accurate peacetime removal rate data and to extrapolate values to wartime programs. The data, however, are historical and may not accurately predict pod performance 10 years from now. Although we have not seen increased failure rates in the last eight years of operation, we also have no indication that this "flat" trend will continue. More important, data collected in the AWOS indicate significantly higher wartime removal rates than predicted using current algorithms.

Figure F.1 shows the removal rates for F-16s flying from Aviano during the AWOS as a function of average sortie(s) duration. Our conservative estimates predict wartime removal rates close to those

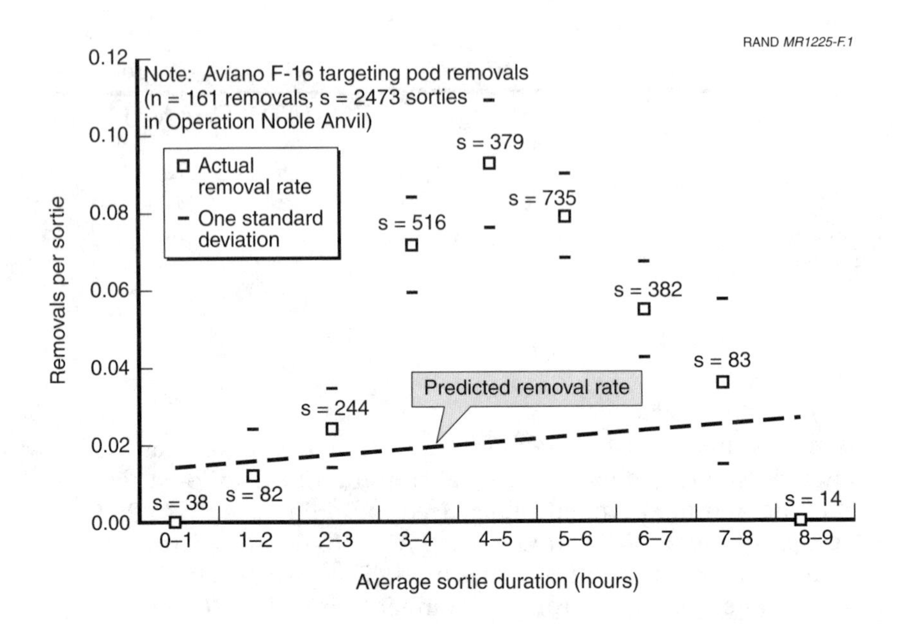

Figure F.1—F-16 Targeting Pod Removal Rates During the AWOS

observed in peace operations (average sortie duration = 1–2 hours). However, the wartime rates are much higher. We modeled this effect through the simulation approach described earlier and included the recomputed removal rates for F-15Es, which varied by well over 150 percent, as shown in Figure F.2.

Simulating the removal rate as well as repair time distributions and expected support equipment processing time improvements generates a 90-percent confidence interval for the expected number of test sets (discussed in the body of the report). In Figure F.3, we show the sensitivity of this output to the expected support equipment performance range described in Appendix B. Note that at the 90th-percentile level there is some effect on tester requirements as a function of expected availability. Yet, as the chart on the right of Figure F.3 indicates, test set resource requirements are very sensitive to expected repair times, implying that future upgrades to the support equipment must be closely monitored to ensure that expected performance levels are achieved.

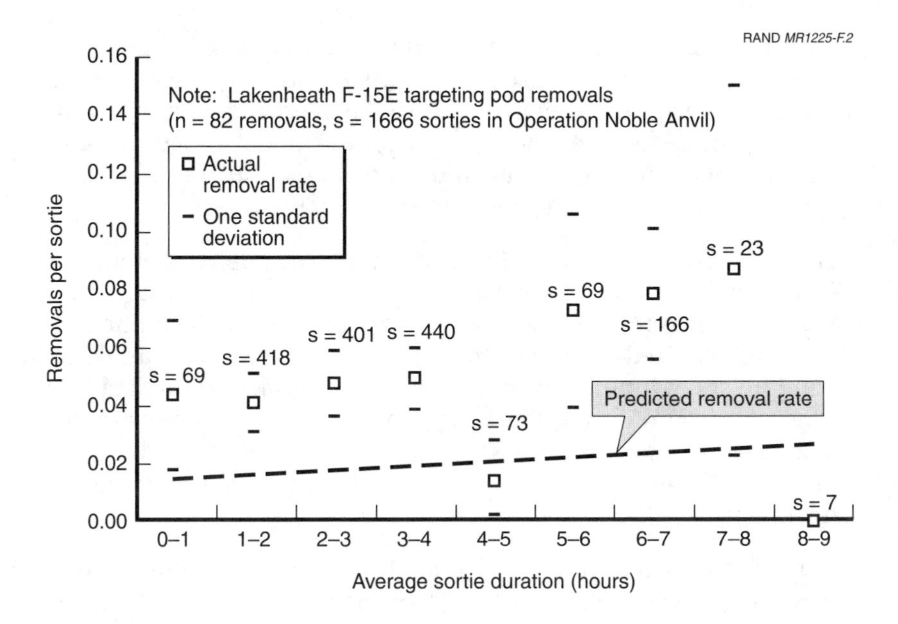

Figure F.2—F-15E Targeting Pod Removal Rates During the AWOS

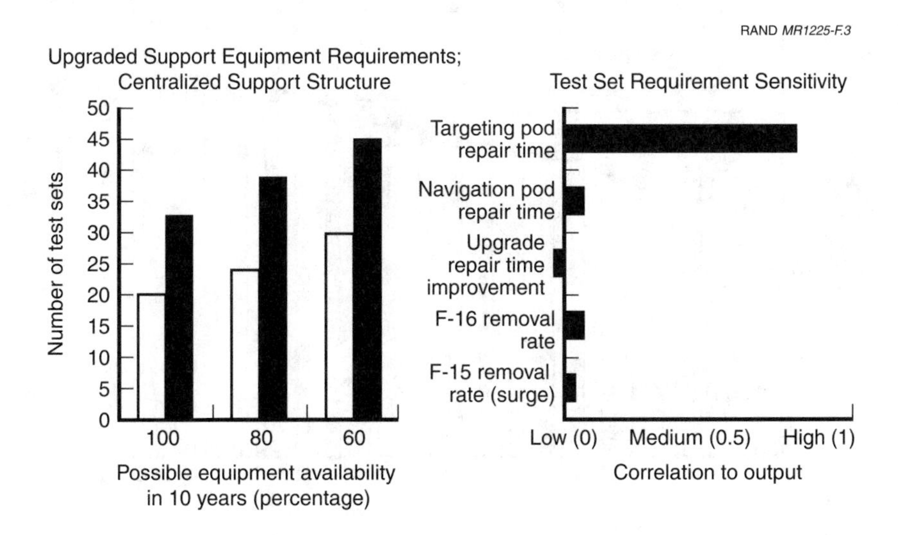

Figure F.3—Sensitivity of Support Equipment Requirement to Availability and Other Uncertain Input Variables

However, one cannot lose sight of the tremendous uncertainty associated with wartime removal rates. As Figure F.4 indicates, a 10 percent removal rate change may drive an average requirement increase of one test set and eight people for a regional structure. And, as discussed earlier, AWOS data indicate that wartime removal rates may be over 150 percent higher than currently predicted.

Personnel requirements exhibit similar patterns relative to removal rates, yet with a stronger sensitivity to expected removal rates and repair time. Figure F.5 shows the relative sensitivity of the number of personnel required versus six uncertainties modeled. We simulated a random set of inputs pulled from distributions based on data collected before and during the AWOS. Again, wartime removal rates are the primary new data assessed from the AWOS. The chart on the right of the figure shows the repair time distribution used to model targeting pods.

We next review our comparative analysis of the range of options assessed. We show resource level requirements based on data col-

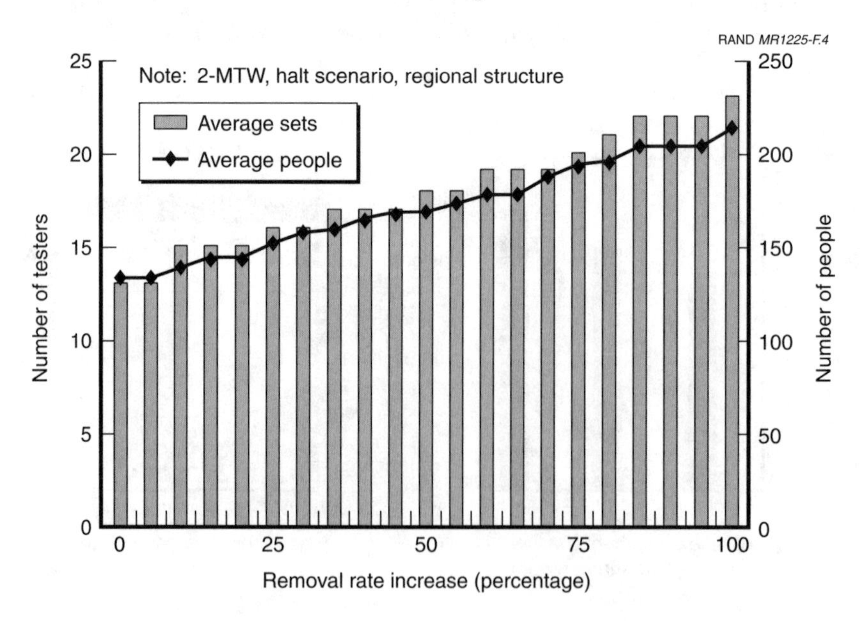

Figure F.4—Test Set and People Sensitivity to Mean Removal Rate Changes

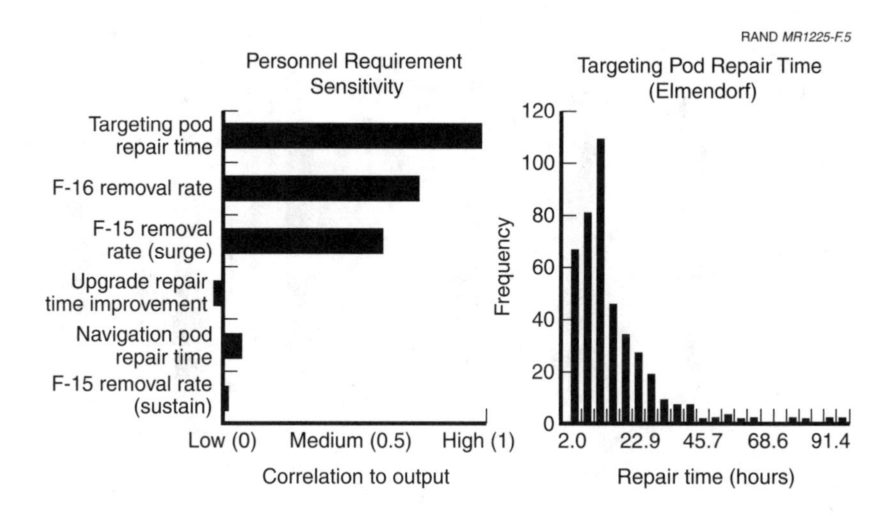

Figure F.5—Personnel Requirement Sensitivity and Targeting Pod Repair Time Distribution

lected during the AWOS. We also depict results based on an assumption of 80 percent availability for single strings regardless of investment level. Earlier we showed that expected availability does not significantly affect our outcomes. Because we address relative value differences, the strategic decision outcomes discussed in the body of the report do not change.

Figure F.6 summarizes the 90th-percentile test set requirement for peace and wartime (halt-phase) scenarios, three logistics structures, and the upgrade option combination. Because the stressing and halt scenarios yield identical resource requirements, we do not show the stressing case results. The white bars represent peacetime needs, and the black bars depict the halt scenario requirements for a two-MTW scenario. We show logistics structures on the horizontal axis, including decentralized support, one CONUS with two OCONUS support locations (three regionals), and a single CONUS location as the lower bound.

The first group of bars on the far left shows the requirement with no support equipment (SE) investment. Note that regardless of centralization level, there are not enough test sets to support two coincident

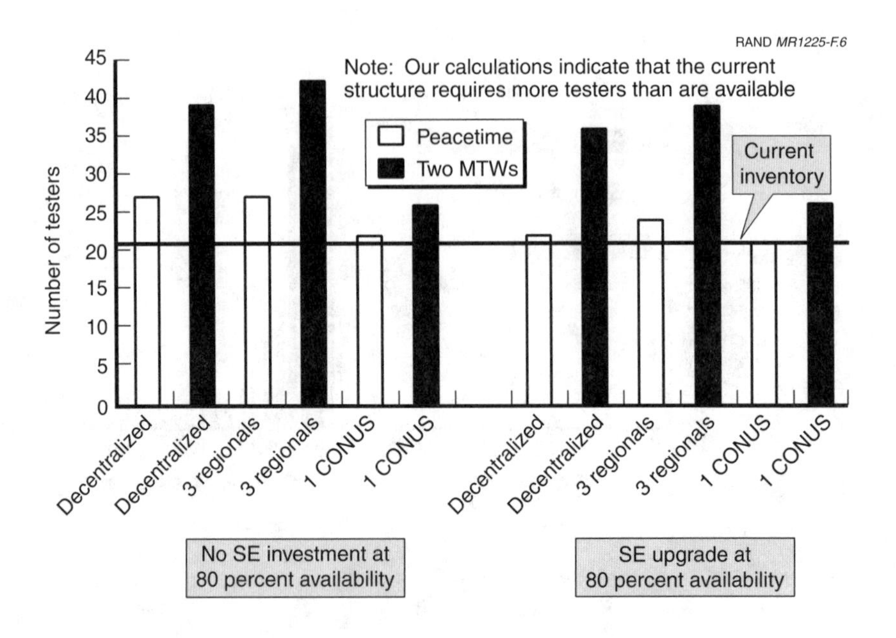

Figure F.6—Tester Requirements by Structure, Investment, and Scenario

MTWs. Although the CONUS-only options offer some opportunities, we showed in the body of the report that transportation capacity limitations prohibit the Air Force from considering this structure as viable. The set of bars on the right show similar requirements for upgraded equipment (we assumed identical availability levels, as discussed earlier). Again, there are too few sets to accommodate coincident wars. Note that a three-regionals option may require more support equipment than a decentralized structure. This finding is primarily driven by potentially slightly oversizing these facilities as a result of our computations at the 90-percent confidence level, resulting in low utilization of the CONUS-based equipment during war and a negation of the collocation effect. Specifically, this option represents a breakpoint in our models where collocation does not reduce resource requirements at the confidence levels we targeted. However, as we move to full consolidation in the CONUS-only option, collocation does reduce the number of required resources.

Therefore, it appears that the Air Force must invest in upgrades to the current system, regardless of whether it continues to use the current decentralized structure or a consolidated structure, if it is to meet the requirements for a two-MTW halt scenario.

From the base of an ADK investment, we calculated two further options, first for the four-echelon structure with ADK for pod removal and replacement and second for the MLU + ADK investment (both not charted). The four-echelon structure, with the ADK investment for pod removal and replacement, requires more ADKs than the decentralized system (with ADKs) because each base would require at least one ADK and then the centralized repair site would require additional test sets. The next level of investment is for the MLU, which includes the ADK plus additional capabilities and performance enhancements. We did not model this option explicitly because we could not predict availability relative to this incremental investment. Figure F.7 displays personnel requirements for the same scenarios,

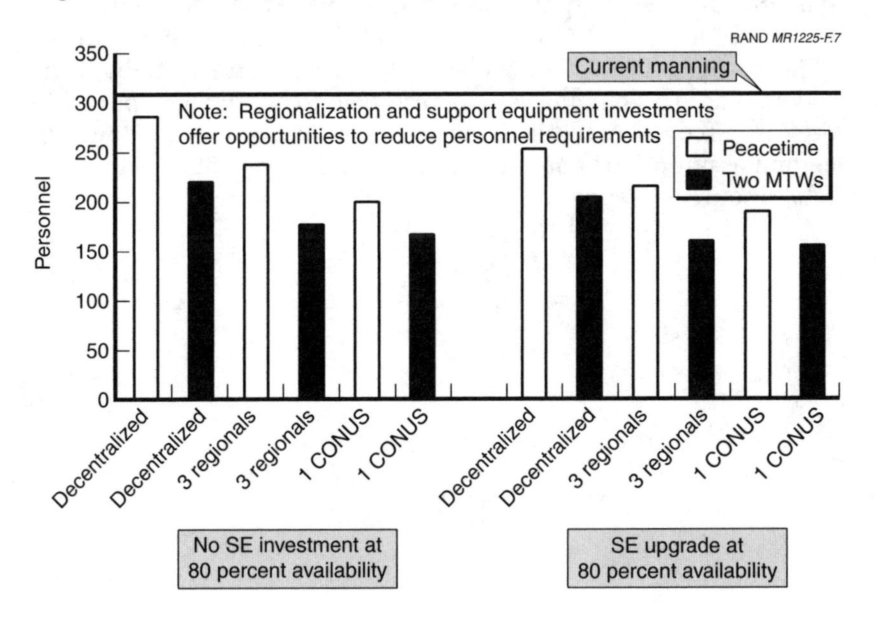

Figure F.7—Personnel Requirements by Structure, Investment, and Scenario

investment, and consolidation options described for the previous graph on tester requirements.

The lower requirement we calculated for wartime is most likely attributable to the manning allocation methodology used in our model. Whereas the Air Force uses a standard back-shop sizing approach to assign LANTIRN support personnel to aircraft, our model employs the industrial engineering concepts described earlier as well as two distinct productivity levels. We also computed the bare minimum manning requirement to support wartime operations (excluding trainees). Conversely, in peacetime we included trainees in the total requirement.

Unlike equipment needs, manning requirements decrease with consolidation. Regional personnel needs were calculated to accommodate AEF rotations into USAFE and PACAF, so the peacetime manning in those regions could be higher than the local peacetime demand dictates. An alternative approach is to keep minimum personnel levels at OCONUS sites and then deploy people to the FSLs in support of AEFs. We show this option in the body of the report. Based on our analysis, there appear to be sufficient people in the Air Force today to fill all contingency requirements. However, certain assumptions about trainees and skill levels may change the total resources needed.

Appendix G

TRANSPORTATION COSTS

To determine the peacetime annual labor and transportation costs of LANTIRN support structures, we investigated several commercial transportation options, including FedEx, DHL, and Emery Worldwide services. DHL cannot support the CONUS routes, and FedEx cannot transport items as large as pods. We therefore focused on the Emery option for pod transport and the FedEx rates for LRU transport.

Our calculations are based on the current USAF tender with Emery and reflect two-day, door-to-door rates. LANTIRN has classified components, so we considered several security measures the USAF could take. The standard rate assumes that the pods are self-insured and no special handling is required. The next higher level of protection is Emery Courier Service. For Courier Service, a third party travels with the pods and keeps a close eye on every step of the delivery process. Constant Surveillance Service is the highest level of protection and is used to transport materials up to the confidential level. Again, the pods are self-insured. A certified and trained person monitors the transport, with a signature trail throughout the delivery process.

Appendix H includes the peacetime transportation costs for pods delivered by the Emery Courier Service.

Appendix H

CASH FLOW CALCULATIONS

Our calculations of present-value costs are derived from cash flow models associated with various investment and recurring costs. Because the AWOS removal rates indicate that the Air Force may be significantly underresourced to meet two coincident MTWs, and because we could not obtain price data for new support equipment (as opposed to the upgrades discussed), we examined the peacetime recurring costs associated with each option.

We first describe the general models developed to assess both investment and operating costs. We discuss only results related to peacetime recurring costs. The labor cash flows could comprise both savings and costs. As investments are implemented over a five-year timeframe (based on Air Force and potential supplier inputs), labor savings may be realized through relocation of personnel not needed to support certain operations. Nevertheless, while potential savings in labor costs typically are factored into commercial-sector capital budgeting calculations, we chose not to include these in our assessment. We did so because the future number of personnel assigned to support LANTIRN will not affect the overall USAF head count, so reduced LANTIRN support labor costs cannot be considered as savings to the USAF.

Thus, it is more instructive to compare actual expected labor costs across the investment options. We estimated an equal incremental increase rate for this cash flow over the investment time period when assessing both regional support infrastructure development and decentralized support. Again, although we discuss our general approach to financial modeling, our results focus on *recurring* (not investment) costs for reasons discussed earlier.

Next, we modeled cash flows associated with transportation requirements for consolidated repair. Again, we assumed that transportation costs would increase during the implementation period and then level off for the remainder of the operational timeline. As with the labor costs, we assumed an equal growth rate in the transportation costs during the implementation period.

Finally, we modeled potential investment costs in terms of costs for LRU spares, infrastructure, and support equipment. We estimated an equal negative cash flow for each year of the investment period, dividing the total expected investment associated with these assets by the number of investment years in this case (or five years based on Air Force and potential supplier inputs). Again, we show only results for the recurring cost computations; regardless of the support option selected, the Air Force may need to make a substantial investment to ensure certain capabilities.

RECURRING OPERATING COSTS

We analyzed peacetime annual operating costs in terms of labor and transportation costs. Labor costs are based on the weighted average of the personnel skill mix factored into our assessment models. We compared Air Force personnel costs with contractor costs (values not shown), so our annual "blue suit" costs include an acceleration factor and additional training costs to account for total costs per person to the Air Force.[1] These costs increase the annual labor costs to the Air Force by about 43 percent above base pay, resulting in a cost equivalent of about $61,500 per person. The option of using Air National Guard personnel was not considered in this analysis.

---

[1]AFI65-503, *U.S. Air Force Cost and Planning Factors,* Table A19-2, A30-1, 1999. The standard rates are a composite and include the following pay elements: basic pay; retired pay accrual (a percentage of basic pay); basic allowance for quarters; variable housing allowance; incentive and special pays that include aircrew, hazardous duty, physicians, dentists, nurses, hostile fire, and duty at certain places. We also include training costs not captured in the composite pay—estimated at about 10 percent of base pay per year. Then, using data in Table A30-1, we apply an acceleration factor (33 percent) recommended for A-76 studies (outsourcing comparisons) for a total of 43 percent above base pay, equaling about $61,500.

We also assumed a certain productivity rate for peacetime operations. This is a metric used in some industries to reflect the number of items produced per paid employee hour. We found that our model outputs are very sensitive to these productivity rates, so we used a value of 60 percent productivity for military manpower during peacetime and 90 percent during war. These numbers also closely match USAF manpower loading factors.[2]

We chose a 2.7 percent discount rate based on the current Office of Management and Budget (OMB) A-94 constant dollar discount rate, and were quoted a rate of about $3 per pound for moving pods with a commercial carrier. Computing military air cost is difficult because of interorganizational financial rules, so we estimated the expected cost per pound for a C-130 transportation network using the Air Materiel Command charge per flying hour of $3839. We then computed the expected peacetime fill rates for the aircraft flying these support loops and amortized the costs accordingly. This calculation yielded a range of about $5 to $8 per pound moved. Although these figures may seem high compared with the commercial option, we note that RAND was given a quote only, with no firm contract, and that commercial carriers may not be able to support certain military contingencies and locations. So again, the decision to centralize hinges primarily on capabilities. Returning to our present-value analysis, we show the relative recurring costs of each option using the commercial carrier transportation rates.

Figure H.1 shows the present value of transportation and labor costs—the two primary operating expenditures—over a 15-year timeline. The consolidated options will carry recurring transportation costs to move parts between support and operating locations. For our analysis of transportation costs, we investigated several commercial transportation options, discussed in Appendix G. Figure H.1 reflects Emery Courier Service (which can accommodate classified items) transportation costs for LANTIRN pods. We do not show our analysis for the four-echelon options (where LRUs are moved) because we found that they do not offer significant benefits in terms of equipment and personnel requirements. Further, we

---

[2]AFI38-201, *Determining Manpower Requirements,* Air Force working team interviews, and site visits.

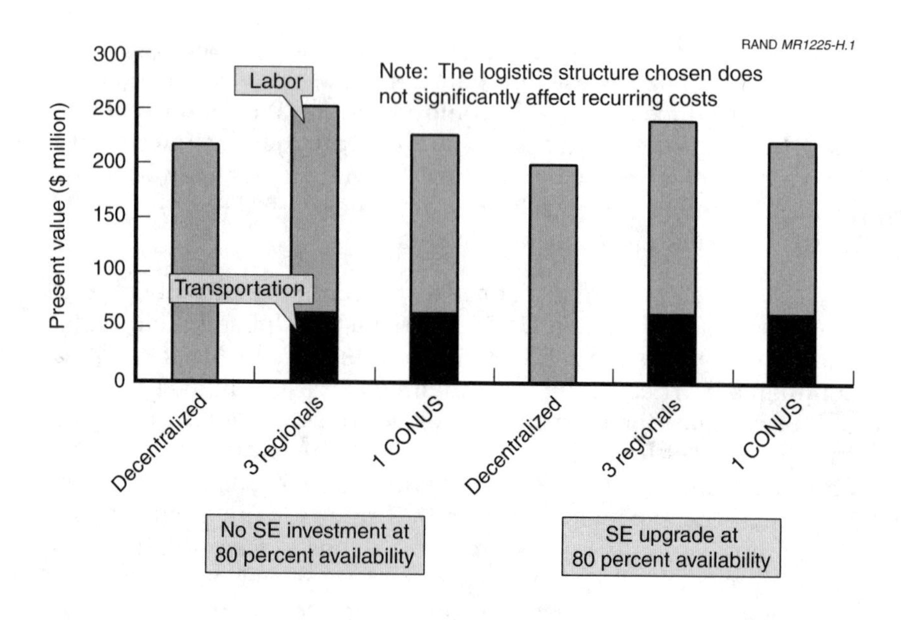

Figure H.1—Present Value of Peacetime Labor and Transportation Costs by Structure, Investment, and Scenario

used proprietary data to compute the contractor labor options, so these results are not shown. We can, however, state that from a financial and operational standpoint, we found no advantage to using contractor labor. Note that there is no marked difference across the six options shown. Although it appears as though single CONUS transportation costs are equal to the three-regional option, we found that commercial rates are actually lower for shipments going to and from the United States, although they may be higher within a foreign region. Thus, we chose to chart the more conservative case, where all transportation costs are equal to those computed for the three-regional structure.

Recurring costs are affected not only by the support structure used but also by the support locations. Establishing regional repair centers where LANTIRN repair already occurs results in lower transportation costs. Seymour Johnson Air Force Base in North Carolina, for example, currently accounts for about 25 percent of the LANTIRN pod assets. Using this base as a proposed regional center could significantly reduce shipping costs. Although we do not show this option in Figure H.1, it warrants consideration in the implementation plans for a regional support structure. Again, the main point here is in the relative comparisons between options. As Figure H.1 indicates, there are no significant operating cost benefits to any particular investment option or logistics structure. Support structure decisions, then, should focus on performance or capability variables and not on costs.

BIBLIOGRAPHY

Galway, Lionel, et al., "Expeditionary Airpower: A Global Infrastructure to Support EAF," *Air Force Journal of Logistics*, Vol. 23, No. 2, 1999.

Hillestad, R. J., and Paul K. Davis, *Resource Allocation for the New Defense Strategy: The DynaRank Decision Support System*, RAND, MR-996-OSD, 1998.

Ochmanek, David, Edward Harshberger, and David Thaler, *To Find, and Not to Yield: How Advances in Information and Firepower Can Transform Theater Warfare*, RAND, MR-958-AF, 1998.

Peltz, Eric, Hyman L. Shulman, Robert S. Tripp, Timothy Ramey, Randy King, and CMSgt John G. Drew, *Supporting Expeditionary Aerospace Forces: An Analysis of F-15 Avionics Options*, RAND, MR-1174-AF, 2000.

Peltz, Eric, et al., "Evaluation of F-15 Avionics Intermediate Maintenance Concepts for Meeting Expeditionary Aerospace Force Support Challenges," *Air Force Journal of Logistics*, Winter 1999.

Ryan, General Michael E., USAF, "Aerospace Expeditionary Force: Better Use of Aerospace Power for the 21st Century," briefing to HQ USAF, Washington, D.C., 1998.

Ryan, General Michael E., USAF, "Air Expeditionary Forces," Department of Defense press briefing, August 4, 1998.

Slay, F. Michael, and Craig C. Sherbrooke, *Predicting Wartime Demand for Aircraft Spares*, Logistics Management Institute, McLean Virginia, AF501MR2, 1997.

Tripp, Robert S., Lionel A. Galway, Paul S. Killingsworth, Eric Peltz, Timothy L. Ramey, and John G. Drew, *Supporting Expeditionary Aerospace Forces: An Integrated Strategic Agile Combat Support Planning Framework*, RAND, MR-1056-AF, 1999.

Tripp, Robert S., et al., "Expeditionary Airpower, Part 2: EAF Strategic Planning," *Air Force Journal of Logistics,* Vol. 23, No. 3, 1999.

Tripp, Robert S., et al., "A Vision for an Evolving Agile Combat Support System," *Air Force Journal of Logistics,* Winter 1999.